HUMAN
GEOGRAPHY
OF THE UK

HUMAN GEOGRAPHY OF THE UK

DANNY DORLING

Cartography by Graham Allsopp

SAGE Publications
London ● Thousand Oaks ● New Delhi

SAGE Publications Ltd
1 Oliver's Yard
55 City Road
London EC1Y 1SP

SAGE Publications Inc.
2455 Teller Road
Thousand Oaks, California 91320

SAGE Publications India Pvt Ltd
B-42, Panchsheel Enclave
Post Box 4109
New Delhi 110 017

British Library Cataloguing in Publication data

A catalogue record for this book is available from the British Library

ISBN 0 7619 4135 5
ISBN 0 7619 4136 3 (pbk)

Library of Congress Control Number 2004094667

*The cover image, being taken from a work of art, obviously excludes parts
of Ireland, the Shetland Isles and other areas outside of mainland UK.
All views of the UK and Britain are necessarily partial.*

Typeset by C&M Digitals (P) Ltd, Chennai, India
Printed on paper from sustainable resources
Printed in Great Britain by The Cromwell Press Ltd, Trowbridge, Wiltshire

For Alison Rachel Dorling

Contents

List of figures and tables

FIGURES

TABLES

Preface

This book has been written for students at university studying at the start of the twenty-first century, assumed to be living in, or interested in the United Kingdom. A social geography and landscape of the country is presented here in a way which is unusual in contemporary teaching.

First, this book contains many maps. Maps fell out of fashion many years ago in the study of human geography. However, the maps in this book are a little different from those which were discarded in the past. Although they initially appear to be showing a crude landscape of the country, the shape and heights of areas on the maps show a social rather than physical landscape. Areas are drawn as mountainous where few children go to university or as lowland valleys where children are most likely to go on to undertake university studies. They are also drawn in size in proportion to the numbers of people being depicted. This, after all, is a book about a *human* geography of the UK.

Secondly, this book uses quantitative evidence. Contemporary statistics from numerous sources have been used to draw the maps and figures shown here. Usually this has required a little analysis of official statistics and in some cases of raw data. In all cases sources and methods are given, in many cases in sufficient detail so that a student can replicate these illustrations through simply having access to the Internet. Just as with mapping, the use of numbers is not currently in vogue in human geography. However, with a little care it is possible to present such evidence in a way which is clear and which reveals many aspects of the underlying geographical structure of society in the UK.

Thirdly, this book is not written as an objective account, although in places it may appear to read that way and the use of maps, figures and quantitative evidence can give that impression. Instead, the book is a story of some of the aspects of life in the UK which are most influenced by people's geographies and a story which begins with ways of imagining the country, childhood and education. It is a story about things which interest me and which I think affect most people, thus geographies of identity, ideology and inequality appear at the heart of the book. It progresses by connecting the geographies of inequality, mortality, work and settlement to these, and ends where it began with a view of children's lives, but of how the children of the UK fit within a global picture of human geography. This is a necessarily partial, parochial, and particular story of the human geography of a country.

All books of the people of a country are written from very personal viewpoints, sometimes all that differs are the extents to which authors admit to that. This book thus has a huge number of omissions. Those which I most regret are the absence of how power and privilege combine in the UK; there are no accounts of individuals' lives and how they are played out on the landscape being drawn; there are no pictures of people or the places in which they live; and a glaring omission of information on Northern Ireland in all chapters save the eighth. If I were starting this project again those are some of the things I would try to do differently.

What this book does hopefully achieve is, at the very least, a description of the country which has not been presented in this way before. A description painted from numbers collected to record key moments in people's lives: their births, movements, literacy, exam results, how they are labelled by the state, how their voices are counted within its democracy, their incomes, expenditures, work, caring, deaths, homes and how these look in a very narrow global context (worldwide childhood poverty). The text accompanying these descriptions suggests something of the processes that have created these images. Many of the processes should be obvious to the people reading this book as they influence their daily lives; they may just be described a little differently here.

For anyone using this book in teaching about the UK, the figures have hopefully been made simple enough to use as illustrations. At the end of each chapter a possible activity is described whereby students can themselves carry out an exercise which illustrates part of what is being suggested in the chapter. Just as you often learn far more by looking at source data than by reading other people's summaries of it, so too it is better to play out what is being described rather than simply listen to such a description in a lecture. These are all exercises I have used in teaching students at ages 16 to 20, from between 12 and 270 in a group. It should not be difficult to interest students in the human geography of a country, especially if they form a part of that geography, but somehow we often manage to turn what should be the most interesting and directly relevant of subjects into an academic exercise in passing exams.

The human geography of the UK is not only of interest to those whose bodies help make it up and can expect to play out most of the remainder of their lives in it. This is a very rich country, as is made abundantly evident in the final chapter of the book. The key question to ask throughout this book is why, given the resources that we have, do we organise ourselves across the country in this way? Why do we have children where we do at the ages that we do? Why do we sort children out both through space and education as shown here? Why do we label people as we do? Why are most of their votes wasted (if they use them)? Why do so many live in poverty? Why can so many not read and write in such a rich country? Why do we tolerate inequalities in illness and death which are so clear to see on a map? Why have we allowed our successful industries to continue to become as geographically concentrated as they have? And why are we following those industries so that many of us are squeezed into very little space while others watch their areas empty out? Before you can ask why, you need to know what has happened, to whom, when, and where. You need to see the human geography before trying to understand it.

Acknowledgements

All faults in the approach, text, figures and statistics presented in this book lie with me. I am most grateful to Graham Allsopp for agreeing to undertake the cartographic work needed to produce the maps shown here which was well beyond my abilities. In this he was assisted by Paul Coles, cartographer at the University of Sheffield, to whom thanks are also due.

Jan Rigby, David Dorling and Dimitris Ballas commented on various drafts of the manuscript, helped turn my elementary English into something a little more readable and to tone down my most bizarre suggestions. I am particularly grateful to Jan for her ability to spot a great many numerous inadvertent *double entendres* and *faux pas*. Any remaining are obviously ones she enjoyed too much to bring to my attention.

Robert Rojek, David Mainwaring and Vanessa Harwood of Sage Publications were all extremely patient in waiting for the manuscript and persevered with remarkable good humour in coaxing it out of me over a two-and-half-year period of my broken promises. The idea for a book of this kind should also be accredited to Robert, who may well wish he'd picked an author who then had not chosen to move work and home yet again, and become a father of two children over the course of this project!

I am grateful to my colleagues at the Universities of Bristol, Leeds and Sheffield for their support when I first and later taught on some of the subjects in these pages; to many of the geography students of those institutions in recent years who kindly endured my experiments in trying to learn how to become a lecturer; and to friends at many conferences and meetings over the last few years who commented on many of the ideas and images that have now found their way into these pages.

Finally I must express my gratitude to the various companies who took over upon the privatisation of British Rail. The excellent way in which they have run the train network of the country since then was largely responsible for giving me the often unplanned time to contemplate what kind of human geography made up this country. In recent years that contemplative service improved greatly. After all, what else is there to think about when staring out of a window wondering exactly where you are, who lives in those houses, how they ended up living there, what they do, and when you might be moving again.

Danny Dorling, somewhere near Trent Junction

1 Maps

...a different view of the United Kingdom

This chapter suggests that your view of this country has been created by the environment in which you grew up. Throughout this book the reader is assumed to be an 18 year-old student who has just gone to university or is just about to go there. Your view of this country has been built up through how the media have depicted the UK, through how you were taught at school to view human geography, and through what your friends and family told you. All these influences on your knowledge were in turn influenced by other events.

The media need to sell newspapers or gain viewers and listeners. They tend to present a salacious view of life in Britain, concentrating on the highs and lows, on the lives of the very rich and the travails of the very poor. The media are fed with government reports, analysis from think-thanks and a myriad of other such sources. Ultimately the media are just another collection of people, most of whom were very like you when they were 18. Most journalists working now went to university. The same is true of your teachers. The bulk of what they will have learnt about the subjects they taught you they were themselves taught in just a few years at university – but sometimes many years before you. What is taught at universities changes over time and is increasingly varied as universities grow.

One thing you can be sure of is that the picture of the UK presented in this short book is not the picture that was taught a decade or two ago. However, this book too is as influenced by its author's exposure to the media, school and the views of his family and friends as your views are. The main difference between my view and yours is that I have had a far greater time to think about the United Kingdom and the people who live in it. One result of this is that the view of the UK that I'll present to you here is a little different from that which you might be used to. It would be a pity if it were not, as then I would not have managed to use what I had learned to try to create some new knowledge.

What I am asking you to do is to bear with me while I try to present you with a different view of the country you grew up in. I am trying to present that view from your point of view. This chapter begins by looking at where 18 year olds live, which of them go to university and from where? How uneven is this landscape, and how in turn is that likely to be shaping the view of the country which university students have? To do this I have to use some numbers and draw some maps. I try to keep both the numbers and the maps simple.

We begin with a question: From where are you looking at the UK? To start to answer that question we need a simple picture of the country. The map we'll take is one that has never been drawn before. It is a crude map, but despite its simplicity it can be used to unveil a great deal about the human geography of this country, its people and, first, its 18 year olds.

You are used to a particular map of the United Kingdom. This is the map you grew up with, the one used in most textbooks and which appears on television every evening in the weather reports, the map which shows the country as it appears from space. However, looking at the United Kingdom from space is not the best way to see its human geography. More people live in London than Scotland, for instance. The alternative map of the UK, shown in Figure 1.1, presents a picture which tries to give the people of the UK fairer representation and which allows us to see variations within large cities alongside variations between regions and between more rural areas simultaneously. The map is of the 85 constituencies drawn up in 1999 for the European parliamentary elections of that year. Northern Ireland was defined as one large constituency that would return three members of the parliament. At the last minute the UK government chose a different voting system for that election and so these areas were not used in that election. We use them here as they present a way of grouping the population of the UK into large adjacent areas, each containing roughly the same number of people. Each is given equal prominence on the map (although some are a little taller than others).

While you may not be used to the map shown in Figure 1.1, the names of the areas on that map listed in Table 1.1 should hopefully be a little more familiar. These are the labels for the 85 constituencies used in this book. Most are named after old counties or parts of counties. They were designed to each contain roughly half a million electors (people aged 18 or over) and to combine

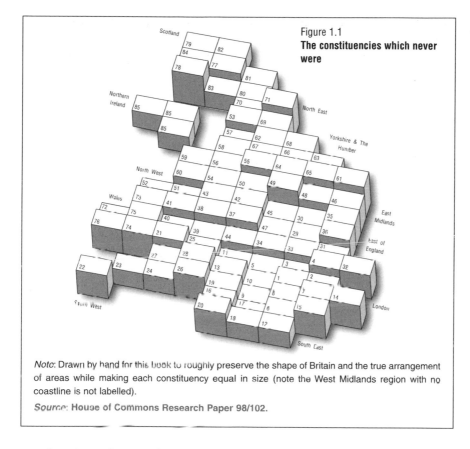

Figure 1.1
The constituencies which never were

Note: Drawn by hand for this book to roughly preserve the shape of Britain and the true arrangement of areas while making each constituency equal in size (note the West Midlands region with no coastline is not labelled).

Source: House of Commons Research Paper 98/102.

together those electors who had most in common geographically (although see the exercise at the end of Chapter 5 to ascertain the veracity of such a claim).

Use the list of names to identify in which constituencies you have lived. If you have difficulty doing this, the full list of which areas make up each constituency is given in the Appendix at the end of this book. Once you have identified your constituency you can see where, on this new map of the UK, you have lived.

Figure 1.1 shows not only each area of the country drawn roughly in proportion to the size of its population, but also gives each area a height. Had this not been done, the map would look more like Figure 1.2. The disadvantage of showing a topography (surface) is that some areas can be slightly obscured behind others which then appear more prominent. An angle of view also has to be chosen and that, too, influences what is seen. The advantage of showing

Table 1.1 Areas that never existed – UK European constituencies, 1999

1	London Central	37	Birmingham East
2	London East	38	Birmingham West
3	London North	39	Coventry & North Warwickshire
4	London North East	40	Herefordshire & Shropshire
5	London North West	41	Midlands West
6	London South & Surrey East	42	Staffordshire East & Derby
7	London South East	43	Staffordshire West & Congleton
8	London South Inner	44	Worcestershire & South Warwickshire
9	London South West	45	Leicester
10	London West	46	Lincolnshire
11	Buckinghamshire & Oxfordshire East	47	Northamptonshire & Blaby
12	East Sussex & Kent South	48	Nottingham & Leicestershire North West
13	Hampshire North & Oxford		
14	Kent East	49	Nottinghamshire North & Chesterfield
15	Kent West		
16	South Downs West	50	Peak District
17	Surrey	51	Cheshire East
18	Sussex West	52	Cheshire West & Wirral
19	Thames Valley	53	Cumbria & Lancashire North
20	Wight & Hampshire South	54	Greater Manchester Central
21	Bristol	55	Greater Manchester East
22	Cornwall & West Plymouth	56	Greater Manchester West
23	Devon & East Plymouth	57	Lancashire Central
24	Dorset & East Devon	58	Lancashire South
25	Gloucestershire	59	Merseyside East & Wigan
26	Itchen, Test & Avon	60	Merseyside West
27	Somerset & North Devon	61	East Yorkshire & North Lincolnshire
28	Wiltshire North & Bath	62	Leeds
29	Bedfordshire & Milton Keynes	63	North Yorkshire
30	Cambridgeshire	64	Sheffield
31	Essex North & Suffolk South	65	Yorkshire South
32	Essex South	66	Yorkshire South West
33	Essex West & Hertfordshire East	67	Yorkshire West
34	Hertfordshire	68	Cleveland & Richmond
35	Norfolk	69	Durham
36	Suffolk & South West Norfolk	70	Northumbria

(Continued)

HUMAN GEOGRAPHY OF THE UK

Table 1.1 (Continued)

71	Tyne & Wear		77	Central Scotland		
72	Mid & West Wales		78	Glasgow		
73	North Wales		79	Highlands & Islands		
74	South Wales Central	**Wales**	80	Lothian		**Scotland**
75	South Wales East		81	Mid Scotland & Fife		
76	South Wales West		82	North East Scotland		
			83	South of Scotland		
			84	West of Scotland		
			85	Ulster (3 seats)		**N. Ireland**

Source. House of Commons Research Paper 98/102 (map on page 49).
Available at: http://www.parliament.uk/commons/lib/research/rp98/rp98-102.pdf

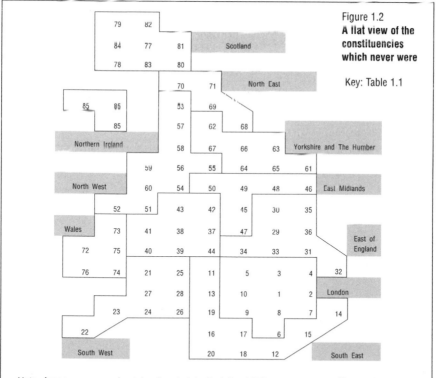

Figure 1.2
A flat view of the constituencies which never were

Key: Table 1.1

Note: Areas are approximately allocated to their best fitting government office region on this figure in England. The West Midlands region with no coastline is not labelled.

Source: As for Figure 1.1.

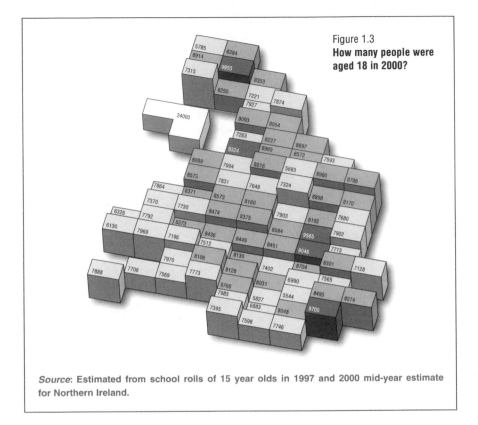

Figure 1.3
How many people were aged 18 in 2000?

Source: Estimated from school rolls of 15 year olds in 1997 and 2000 mid-year estimate for Northern Ireland.

a topography is that it is always possible to view what is being mapped in relation to another variable depicted by the height of each area. In physical geography it is height itself which is usually depicted – rivers run down mountains, temperature tends to fall as the land rises and so on. In human geography there is no single obvious variable to use to map the basic contours of the social landscape. However many social variables produce very similar landscapes and so the precise choice is not critical. Here I have taken the first life chances measured in this book (in Figure 1.5) as this should be of interest to the reader. Height on all the maps drawn here is in proportion to a child's chances of not winning a place to attend university. These chances have been turned into a categorical variable to simplify the landscape. The higher an area appears, the fewer people growing up there go on to university.

This book is being written for the hypothetical university student who turned 18 in the first decade of the millennium. Where might those students have

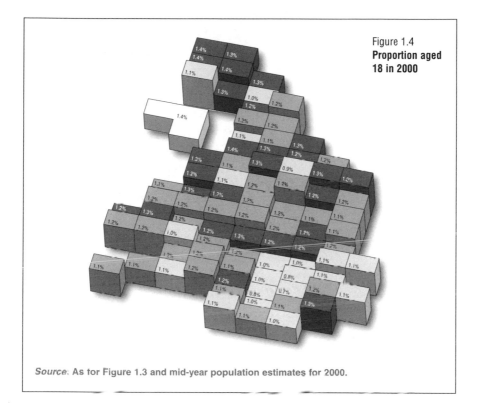

Figure 1.4
**Proportion aged
18 in 2000**

Source: As for Figure 1.3 and mid-year population estimates for 2000.

come from? Figure 1.3 provides an answer based on how many 15 year olds attended schools in each constituency three years earlier. For Northern Ireland these statistics were not available and so the map uses the much rougher official estimate of 18 year olds living there based on births, deaths and migration records. Across Britain the map shows disparities of the order of many thousands in the number of people who turned 18 in each area of Britain in 2000. Part of the pattern will be due to the fact that each constituency was designed to contain roughly half a million voters, not exactly equal numbers of people (there were just over a million voters in the three Northern Ireland seats). To see the degree to which there is actual variation in this population group we need to divide the numbers counted for Figure 1.3 by the total population living in each constituency.

Figure 1.4 shows the proportion of 18 year olds based on their estimated constituency of origin (where they were at school when aged 15) as a proportion

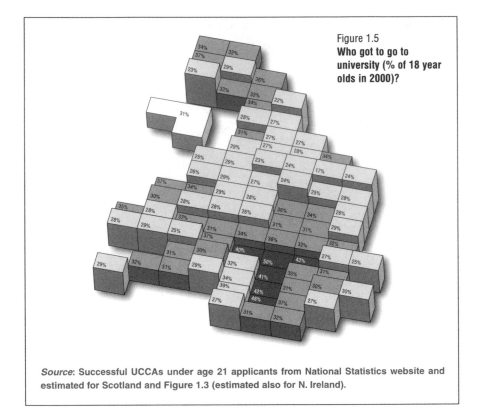

Figure 1.5
Who got to go to university (% of 18 year olds in 2000)?

Source: Successful UCCAs under age 21 applicants from National Statistics website and estimated for Scotland and Figure 1.3 (estimated also for N. Ireland).

of the total population living in each constituency. The pattern it shows is of low proportions of 18 year olds living around the south coast, in London and in the centres of some other large cities. The three areas with the lowest proportions were London South Inner, London Central and London South West, followed by Sheffield. What led to less than 1% of the population being aged 18 in these places?

Eighteen year olds in the population of the UK are most likely to come from the more suburban constituencies of the UK. The highest proportions are found in Lancashire South, followed by the West, Centre and Highlands of Scotland then by Cheshire East, Cleveland & Richmond (North Yorkshire) and Yorkshire West. By the time their children are aged 15, a significant proportion of parents have moved their family homes away from the city centres. Later we'll see how many were born there and more about this selective migration.

Table 1.2 The two extreme European constituencies compared

London North West	UCAS	School	%	Yorkshire South	UCAS	School	%
Harrow East	765	1142	67	Don Valley	280	1105	25
Brent East	360	541	67	Rotherham	215	897	24
Brent North	590	978	60	Doncaster Central	250	1165	21
Hayes and Harlington	265	499	53	Doncaster North	165	802	21
Brent South	400	790	51	Barnsley Central	155	1031	15
Harrow West	650	1340	49	Wentworth	185	1324	14
Ruislip–Northwood	400	956	42	Barnsley East & Mexborough	165	1396	12
Uxbridge	280	1156	24	Rother Valley	140	1240	11
Total	3710	7402	50	Total	1555	8960	17

Note: UCAS = successful applicants; School = number of pupils aged 15.
Source: As for Figure 1.4.

One reason for the drift of children to the suburbs could perhaps be that some parent(s) are attempting to improve their offspring's life chances by moving home. Figure 1.5 shows the proportion (17% to 50%) of 18 year olds who enter higher education from each constituency. In general, a higher proportion enter from the suburban and southern constituencies (although entry from much of Scotland and North Wales is above average). However, the children with the highest chances of going to university went to school in a city constituency: London North West (50.1% entry), closely followed by more rural Surrey (47.7%). The constituency where children's chances of going to university was lowest was Yorkshire South (17.4%) where children were almost three times less likely to go to university than in London North West. It is worth noting that the estimate for Northern Ireland is almost certainly an underestimate as it does not include participation of students from Northern Ireland in universities in the Republic of Ireland. What was your chance of going to university? Note that by the time this book is published most people's chances will have risen.

Statistics about people should never be taken at face value. Table 1.2 shows the numbers which were used to calculate the statistics for the most extreme two areas shown in Figure 1.5. The European constituencies of both London North West and Yorkshire South are defined as a combination of eight UK parliamentary constituencies. Within each European constituency there is a great deal of geographical variation. For instance, taking the extreme two parliamentary constituencies you could claim that children in Harrow East were six times more likely to go to university than children growing up in the Rother Valley. If you wanted to downplay the differences, you could point out that children in Uxbridge were less likely to go to university than children in the Don Valley. But how reliable are these figures?

Although there may have been 1,156 children aged 15 attending schools in Uxbridge in 1997, more of those children probably lived outside that parliamentary constituency than commuted out of it to go to school. The UCAS figures for entry to university are based on people's home address, not their school's address. We are therefore not strictly comparing like with like, and this is before we start to worry about the migration of children between age 15 and when they apply to university and so on. At the level of the European constituencies such problems are less acute because a far lower proportion of children will cross European constituency boundaries to go to school. Nevertheless it always pays to think about where the data comes from that is presented to you as fact.

The UCAS data that has been used so far in this chapter is provided on the government's official website for every local government ward in England (http://www.neighbourhood.statistics.gov.uk/home.asp). If we knew a reliable count of the population eligible to apply to university from each ward, we could produce statistics that showed even starker geographical differences than those presented so far. If you went further still and had data on individual 18 year olds, then the differences between who did and who did not go to university could be made to appear even starker between individual streets and houses. However, at the broad geographical level, between areas that contain hundreds of thousands of people, the differences will have more to do with the areas in which people live than with the individuals themselves.

If statistics within one country are difficult to interpret, check and understand, the statistics which compare rates between countries are even more problematic. Figure 1.6 presents participation rates for 18 year olds in secondary education for 14 Western European countries. The rate for the UK (at 32%) is very low in comparison with most other countries and surprisingly similar to the rate entering universities some four years after this data was collected. Note that this is the 'full-time' participation rate and note when the

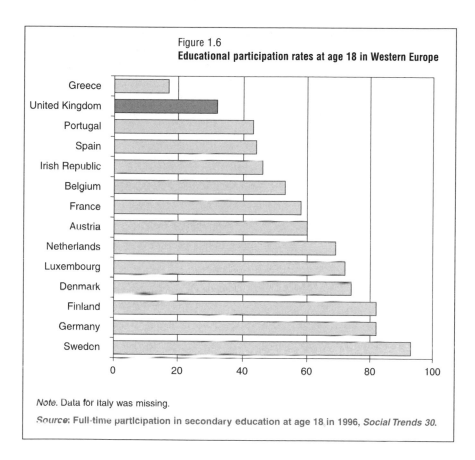

Figure 1.6
Educational participation rates at age 18 in Western Europe

Note. Data for Italy was missing.

Source: Full-time participation in secondary education at age 18 in 1996, Social Trends 30.

data was collected. A large part of the reason for the overall level of university entry in the UK is due to a traditionally low level of staying on at school past the compulsory age of 16. For instance, in Belgium, The Netherlands and Germany the school leaving age is 18. When trying to estimate a child's chances of going to university what matters most is that they grew up in the UK. The next most important factor to be determined is where they grew up in the country.

How do we know that location is so important in determining an 18 year old's chances of going to university? Surely it could just be that children growing up in better-off families are more likely to go and those families are concentrated in particular places? Well, the advantage of counting these things is that we can check these ideas.

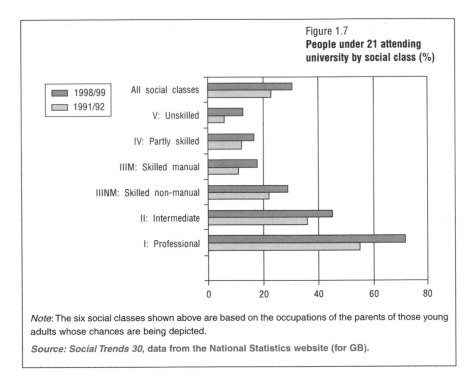

Figure 1.7
People under 21 attending university by social class (%)

Legend: ■ 1998/99　□ 1991/92

Note: The six social classes shown above are based on the occupations of the parents of those young adults whose chances are being depicted.

Source: Social Trends 30, data from the National Statistics website (for GB).

Figure 1.7 shows the increasing proportions of students attending university according to the occupations of their parents. The distribution is very uneven but has become slightly more even over recent years. By the turn of the millennium 72% of the children of parents in professional occupations were going to university as compared to 13% of the children of parents in occupations labelled as unskilled. Looking at this graph you might be led to believe that it is these differences that account for the geography of participation shown in Figure 1.5 earlier. How do we go about checking that?

For each European constituency we know the proportion of children (aged 0–15) of parents of different occupations as recorded in the 1991 census. Those children had an average age of between 7 and 8 in 1991 and so are roughly representative of our 18 year-old adults in 2000. The social profile of areas also tends to change only slowly over time. Areas with lots of people in professional occupations in one year tend to have lots of people in those occupations in the next year. Thus the social profile of children by area by the class of their parents can be expected to have been fairly constant since 1991.

Table 1.3 Children by social class and predicted university entry rates

children by social class	I	II	IIINM	IIIM	IV	V	Total
London North West	9%	34%	15%	30%	9%	3%	100%
Yorkshire South	4%	22%	9%	40%	18%	7%	100%
GB % going to university	72%	45%	29%	18%	17%	13%	(of total)
predicted proportions							
London North West	7%	15%	4%	5%	2%	0%	33%
Yorkshire South	3%	10%	3%	7%	3%	1%	27%
actual proportions							
London North West							50%
Yorkshire South							17%

Source: 1991 Census 10% statistics and Figure 1.7, 1998/99 (for class labels).

So given that we know the proportion of children in each social class in each area and the proportion of children from each social class who go to university, what would we expect the proportion of children going to universities from any given area to be?

For our two extreme constituencies Table 1.3 presents the proportions of children in each social class using the same labels as in Figure 1.7. More than twice as many children have parents in social class I (9% professional) in London North West when compared to Yorkshire South (4%). More than twice as many children in Yorkshire South have parents in unskilled occupations (7%) as in London North West (3%). The two constituencies look to have very different social profiles. In a way they do, at the extremes, but in each roughly three-quarters of all children have parents in social classes II, IIINM and IIIM (combined). How might these differing constituency social profiles be expected to result in differing numbers of children going to university from these two places?

The mathematics for a prediction can be very simple. Here we just take the national proportion of children going to university in each social class and multiply that by the proportion of children in that class in each place to produce a prediction of what proportion of children of each class in each place we would expect to be going to university. Thus we would expect 72% of the 9% of children in social class I in London North West to go to university: 72% of 9% = 7%. Adding up all the predicted proportions along the columns for

London North West results in an overall prediction that 33% of children there will go to university. Doing exactly the same procedure for Yorkshire South results in a predicted proportion going to university of 27%.

We know that the actual proportions of 18 year olds going to university from the two extreme constituencies is 50% and 17%, which are 17% higher and 10% lower than we might predict. To put it another way, allowing for the differing social class profiles of these areas only helps to explain 6% (33% – 27%) of the 33% (50% – 17%) variation between them, or about one-sixth of the geographical variation. This is because although there are very great differences in university entry rates between the chances of children whose parents have different occupations, the distribution of children by social class does not vary that greatly geographically. At least it does not vary that greatly between these areas that each contain roughly half a million electors. Each of these large areas has its share of children from better-off and worse-off families. The differing social profiles of these areas only explain a minor part of the differing levels of university entrance (at least for the extreme areas).

Figure 1.8 presents the results of subtracting the proportion of children we might expect to go to university in each area from the proportion who actually do go. The largest positive discrepancy is London North West at 17%. Other areas where more children go than might be predicted include Surrey (+9%), London West (+8%), London North (+7%) and Leicester (+6%). Areas where fewer go than predicted are headed by Yorkshire South (rounded to –9% on this map), Bristol (–7%), Essex South (–6%), Sheffield (–6%) and Nottinghamshire & Leicestershire North West (–6%). The map is shaded by those areas where more than the expected proportion go to university. Note that percentages near zero are rounded to zero but the areas are shaded dark if they are positive. There are still significant variations between areas, even when the differences are this slight. Social class only explains a little of the map of university entry.

What could account for the remaining variation between these areas? The answer is likely to be many different things. There are often no simple single explanations to geographical patterns and this is part of what makes the human geography of the UK so interesting. Analysing statistics such as these allows new ideas to be generated on the basis of current knowledge which in turn can be analysed themselves until there is little left to try to explain. For instance, the desire to leave the most remote parts of Scotland, North Wales, Northumberland and North Yorkshire could drive children to try particularly hard to get to university. The schools in other areas might help their children

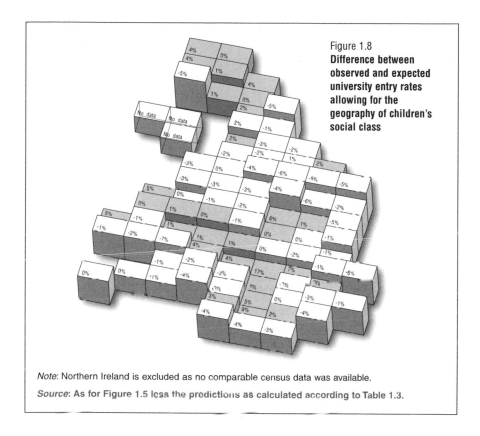

Figure 1.8
Difference between observed and expected university entry rates allowing for the geography of children's social class

Note: Northern Ireland is excluded as no comparable census data was available.

Source: As for Figure 1.5 less the predictions as calculated according to Table 1.3.

to achieve lower than average marks in their exams (we look at GCSE exam results by these areas later). There are more university places in the south and centre of England and so more children from areas in this central belt might apply to universities as they are nearer home. There are many possible explanations.

The most important point that this introductory example makes is that it is not possible to reduce something as superficially simple as university entry into a purely social process whereby different social groups experience different chances and those social differences are what the map of university entry rates is portraying. There is much more to the map of 18 year olds' chances of attending university than that. That extra information, the majority of the differences between areas, is just a very small part of the overall human geography of the UK.

If you want a simple definition of what human geography is, then it is the study of the spatial variation in the lives of people that cannot be reduced to purely social, political or economic processes. Human geography attempts to understand and explain people's lives in the round. Our concern is about where people are, where they have been and where they are going.

An exercise
(6 to 600 players)

You do not need a computer, or even a piece of paper and pen, to draw a map. Human bodies themselves will suffice. Here is a recipe for drawing a map of the geographical origins of a room of students. 'Cooking time' is about 10 minutes, although allow slightly longer when undertaking this exercise with over 100 students.

1   Each student needs to decide where they 'come from' (were born), i.e. their home city and suburb. If a student comes from outside the UK, that is fine – they will simply be appended to one of the edges of the map we are about to draw.
2   Determine the four corners of your map. Who comes from the furthest North East, North West, South East and South West? These four students need to move to stand at the top right, top left, bottom right and bottom left of the classroom/lecture theatre respectively.
3   Now every other student must begin to sort themselves out in the map of where they come from. Begin by sorting yourselves out from North to South. The further North you come from, the further back in the room you need to be. Ask your neighbours directly in front and behind you to work out if you are in the wrong place.
4   Now sort yourselves out from East to West. The further East you are from, the further right you should be along the row you are in. Again, by asking your neighbours (either side of you) where they are from, you should be able to work out if you are in the wrong place.
5   Now check with all four of your nearest neighbours (those to your left and right and those in front of and behind you) to find out if they come from further East of you, further West, further South and further North of you respectively. If they don't, you need to move around a little more.
6   Finally, all shuffle in towards the centre of the room to end up with just an arm's length between you and your four nearest neighbours. Once you have done this you will have created a map of the country in which the area in the room is arranged in proportion to the population of students in your class by their areas of origin.

Having created your own map of the country, what you next use it for is up to you. If you are undertaking this exercise in an 'old university', your map is likely to include most of the country as people tend to travel further away from their areas of origin. However, what are the geographical biases in your distribution? Where does the person in the centre of your map come from? Further South or North than most people in the country? For reference, the central constituency in Figure 1.1 is the Peak District in Derbyshire. If you are undertaking this exercise in a school, then your map is likely to be only of a very small part of the country – but it is still a map drawn in proportion to you. Does the person in the centre of your map live closest to your school? If not, why not? Next you could begin to look for geographical differences among yourselves. If at university, then sit down if you took a 'gap year' before starting your studies. If at school, sit down if you intend to go to university (or take a gap year). Did more students to the South or North of your particular map of Britain sit down? You could divide the class into groups depending on where they originate from and then poll those groups to see if there are differences on their attitudes, say, to voting. These polls can be done anonymously on scraps of paper. The point is that even within one room there are likely to be geographical patterns. These are patterns which cannot simply be explained by the social, economic or political backgrounds, statuses and beliefs of those being counted.

[Note: This game can easily be played for other countries and regions in the world if students have some link to those. If playing this game in the Southern Hemisphere, it is customary to place South towards the back of the room rather than north. An advanced version of the game, designed to simulate the panic that can ensue following epidemic disease outbreak, involves all students forming the original map as described above, but then trying to move to be as far away from their four geographical neighbours as possible while still remaining in the room. This version is best not played with students under age 18 or classes of more than 60 given the consequences.]

CONCLUSION

It can take a little time to get used to the human geography landscape of the UK if you have not seen it before. Look again at Figure 1.1 and follow this account of its contours (use Table 1.1 and Figure 1.5 as a guide also):

The landscape is lowest, and the ground most fertile for prospective students in a T-shaped valley which encompasses the northern and western suburbs of London stretching out from there to Oxford in the west and to Surrey in the south. Surrounding that valley is a plain only a little less fertile than the valley itself. From central London it reaches down to the Sussex coast, across the South Downs, up the Thames Valley to Hampshire, then across to Gloucestershire, Somerset, Dorset and Devon. The northern border of the plain stretches from Shropshire, through Herefordshire, Warwickshire, Northamptonshire, Leicestershire, Cambridgeshire, Bedfordshire, Suffolk, ending at the Essex coast. Around the plain are found the foothills of more stony ground where life chances are a little less sure for the children of North East and South East London, Kent, the Hampshire coast, Southampton, Wiltshire, Bristol, Cornwall and most of Wales, the Midlands and the North West of England. The land tends to rise as you move northwards. There is a particularly noticeable cliff as you cross into the north of Nottinghamshire and up to the highlands of Greater Manchester (East), into Sheffield, the rest of South Yorkshire and East Yorkshire. Further north the land becomes mountainous again in Tyne and Wear and across most of Scotland, with the highest peak in the UK being Glasgow. Northern Ireland is part of the uplands. There are some dips within the north and west, but these are not extensive enough to form anything approximating the great southern plain and there is no fertile valley in the north. These dips include parts of Mid and West Wales, Cheshire, Central Lancashire, Northumbria, Southern Scotland and Edinburgh.

The human landscape of the UK is not dissimilar to its physical geography. However, even in the north, cities tend to have been built in small valleys whereas in this landscape they can form great mountains. There is no need for water to flow to the sea over a landscape made up of human chances. It is just fortunate from the point of view of engaging the imagination that, if you think of the River Thames running between East and North London, the human geography topology is also a plausible set of physical contours. Although had we the space to draw them, we would not draw rivers on the maps in this book, but motorways, railways, airports and sea crossings. These are the rivers which carry people over the land. However, drawing these would have complicated the maps further. Already they allow you to compare two geographical distributions simultaneously and place numbers and symbols in each area. Given the readership of this book, and to avoid unnecessary confusion, throughout these pages one of those distributions is always held static as height – as the landscape – that distribution being the chance of attending university, dependent on where you grew up.

2 Birth

...and the suburban pied piper

This chapter begins the story of the human geography of the UK at the beginnings of life by starting with the history of how many children are born here each year and where they subsequently moved to before reaching adulthood. The chapter ends by looking at where and when most of these children, as adults, have children themselves. Before that we consider demography's contribution to the patterns uncovered in Chapter 1, which are the products of many other patterns. Most significant among these has been the historical development of education and privilege in the United Kingdom. If you are reading this book as a UK-born university student in the early years of the twenty-first century in this country, then by far the most important factor that resulted in you attending university was the year of your birth. The more recently you were born, the more likely you were to attend university. Before we look at just how likely this was, for many other chances in life, how many other people you were born with also turns out to be of great importance.

Figure 2.1 shows the numbers of babies born each year of the last century in Britain. At the peak, 1.1 million babies were born in 1920, 300,000 more than in 1919. People waited until war was over to have children. Similarly, 150,000 more babies were born in 1946 as compared to 1945. Births peaked again in 1964 and 1990, with intervals of 18 years and then 26 years between these post-war highs as the average age at which women had children rose. This graph does not, of course, tell you what the fertility rates in different years were as there were growing numbers of women over the period. What it does tell you is that if you were a child born since the lowest birth year of 1977, then you belong to a cohort of children which was (last) as small as those born over 50 years ago. There are about 200,000 a year (or more than a fifth) fewer of you than there were in your parents' generation.

Why should the size of your birth cohort matter as compared to that of previous generations? Well, among many other things it will influence the

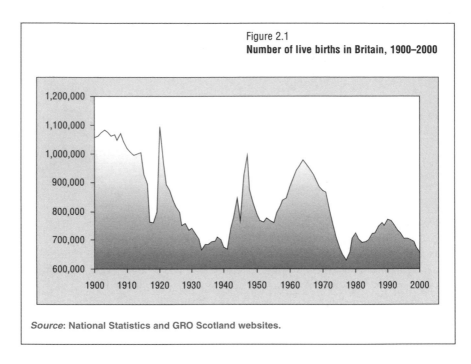

Figure 2.1
Number of live births in Britain, 1900–2000

Source: National Statistics and GRO Scotland websites.

chances and choices you later have over employment, the kind of home you can live in and your educational chances when the numbers of places at university are fixed. As a larger, older generation retires from work there are fewer young people to replace them. As the homes built for the larger families of the past are vacated there are larger houses to move into. But there are also fewer of you to care for a larger and ageing population and fewer of you to have children yourselves, which is one of the reasons that the trend is downwards in the number of births per year in the last few years shown in Figure 2.1. The number of births is predicted to continue to fall until around 2006, then rise slightly until 2020 before falling again to a new low in 2037.

Having said that year of birth matters (given you were born in the United Kingdom), then for many things the next most important determinant of your life chances is whether you were born female or male. These key chances range from how long you might live to what job you might have to, again, your educational chances. Intriguingly, your chances of being born of either sex have also varied over time. Slightly more babies have always been male but, as Figure 2.2 shows, the proportion of female babies has fallen over the course of the century from a high of 491 per thousand in 1900 to a low of 485

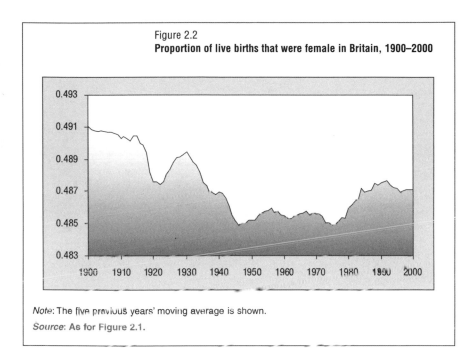

Figure 2.2
Proportion of live births that were female in Britain, 1900–2000

Note: The five previous years' moving average is shown.

Source: As for Figure 2.1.

per thousand in the mid-1970s, rising slightly again to 487 per thousand by the end of the century. It is an interesting question to ask why this should be so. However, for all practical purposes, almost as many women are born as men but, because girls' survival rates are slightly better than boys', by age 21 the numbers of men and women living in the United Kingdom are almost identical.

If we turn to look at the key life chance this book began with – your chance of entering university – the importance of birth cohort size becomes apparent. The first bar chart in Figure 2.3 shows the rise in the number of full-time undergraduate university places in the UK between 1970 and 2000 for men and women; the second shows the numbers of people turning 18, 19 or 20 in that year (estimated from the births occurring some 18, 19 and 20 years earlier); and the third chart shows the first figures as a proportion of the second set as a crude estimate of university entry rates. The simplicity of the estimate should be clear, the birth data does not include Northern Ireland, some children migrate or die before reaching 18 and increasingly undergraduates come from a greater age range than this. However, the apparently smooth growth of the crude numbers of students is revealed to have instead been stagnation in the

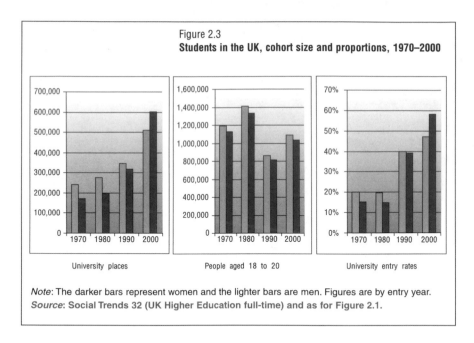

Figure 2.3
Students in the UK, cohort size and proportions, 1970–2000

University places | People aged 18 to 20 | University entry rates

Note: The darker bars represent women and the lighter bars are men. Figures are by entry year.
Source: Social Trends 32 (UK Higher Education full-time) and as for Figure 2.1.

proportion of those born in the early 1950s and 1960s going to university. There followed a doubling of the proportion of men and near tripling in the proportion of women during the 1980s and further rises in the 1990s, with women being a quarter more likely than men to become a full-time undergraduate by 2000.

Thus the cohorts of students are growing at the same time as the cohorts of the population as a whole are generally falling and the proportions becoming students have never been so high. These proportions are higher than those shown in Chapter 1 because overseas and Open University students are included as well as mature students, but part-time and further education students are omitted (which show even greater rises for women). At the start of the century only some 25,000 students attended university, under 1% of the equivalent population and almost all male. The proportion of those students who foresaw that women would out-number them by the end of the century is unknown!

It is perhaps in the years between birth and age 18 that a person's chances in life are largely set. The Jesuits famously said that given a child until he is seven and they could give you the man. If you look back for evidence for this

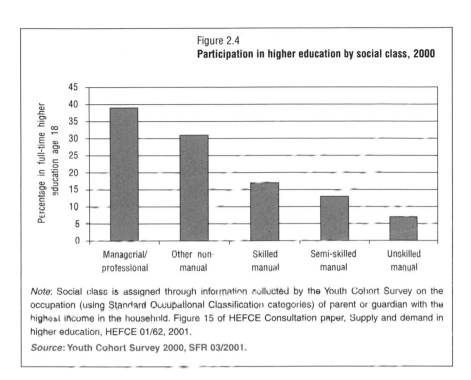

Figure 2.4
Participation in higher education by social class, 2000

Note: Social class is assigned through information collected by the Youth Cohort Survey on the occupation (using Standard Occupational Classification categories) of parent or guardian with the highest income in the household. Figure 15 of HEFCE Consultation paper, Supply and demand in higher education, HEFCE 01/62, 2001.
Source: Youth Cohort Survey 2000, SFR 03/2001.

quotation, you may in fact find that they were actually saying they did not want to educate children before the age of seven! Others argue that life chances are largely set before or by birth, while yet others believe that almost any 18 year old could be privileged from that point onwards and have their chances in life drastically altered. These are largely academic arguments. In practice, as a society we do not tend to intervene in life systematically enough to test these arguments. Other than through adoption, we do not take children at young ages and place them in other families to see what happens to them. Whether a child achieves something such as going to university or not, in Britain then, is largely determined by the family into which they are born (Figure 2.4). However, families are far from idle in this process. An increasing number pay for their children to attend private school to try to ensure things such as their university entry. Others pay for home tutoring, provide additional teaching themselves or bribe their children to 'try harder'. Still more improve their children's chances by moving geographically so that their children go to a supposedly better school. Like cattle moving up to higher slopes for new grass in the summer, a large proportion of the children of Britain are herded into fresh pastures as they age. This migration of school-age children

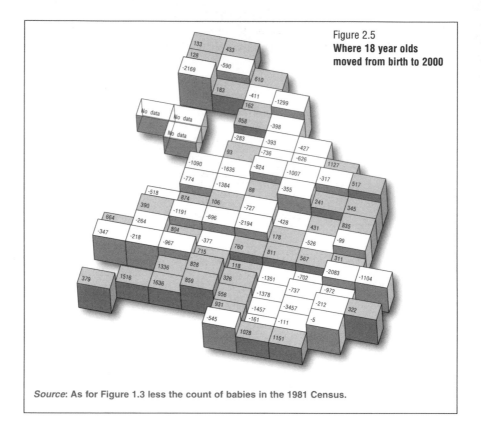

Figure 2.5
Where 18 year olds moved from birth to 2000

Source: As for Figure 1.3 less the count of babies in the 1981 Census.

and their parents is one of the most significant population movements in Britain and so this is where we turn to next.

The children who became adults in 2000 were born in 1982. One source we have to estimate where they were born is the census taken the year before their birth. If we assume that roughly the same number of babies was born in 1982 as in the year up to the census held in spring 1981 and in roughly the same places, then in Britain we begin with some 683,464 babies. To make these estimates we are not just ignoring the difference in timing, but also the migration of children into and outside Britain and the deaths of some of those children before they reached age 18. Nevertheless this estimate compares well with our total estimate of there being 669,269 18 year olds in 2000 (itself based on counts of school children in 1997!). In Figure 2.5 the count of babies estimated to have been born in 1982 is subtracted from the count of 18 year olds living in Britain in 2000. The map suggests that there was indeed

HUMAN GEOGRAPHY OF THE UK

a significant movement out of the cities from birth. This was highest in London South Inner where 9001 babies were born, but only 5544 18 year olds remain: a net outflow of –3457, as shown on the map. Birmingham East and Glasgow recorded the next largest net outflows. The three largest net inflows were in Dorset & Devon East, Devon & East Plymouth and Somerset & North Devon.

The flight to 'better schools' is only one of the factors underlying the movements shown in Figure 2.5, but it is probably the key factor. It is important to remember that Figure 2.5 only shows net change. It hides a huge amount of gross change. In many places (as shown in Figure 2.7 below), for every child that moves out of an area, another child (and their family) moves in to take their place. There are relatively few new homes being built and relatively few (although growing) numbers of areas being abandoned in Britain. The way in which children are sorted by education is largely a zero sum game. More go to university each year and fewer leave schools with no qualifications but the meanings of a degree and having no qualification also alter over time. There are only a fixed number of chairs in the classrooms of the 'better schools', although more of these seats have bums on them than in other schools. By definition, half of all children in Britain have to go to below-average schools. Where education remains a social sorting exercise and school league tables are still published, this cannot be stopped (it has stopped in Wales, although parents in the know will still know).

Figure 2.6 shows how important school exam results are, almost certainly both influencing the directions in which children migrate and how that migration influences the examination results. The degree of variation in the graphs suggests that there is a little more going on here than a very simple geographical relationship (that is the subject of the next chapter). However, several thousand children move out of areas where high numbers are not awarded high grades at age 15 or 16. In net terms, it is into the areas where the majority are awarded high grades that they move to. Hence the areas in the graph in Figure 2.6 tend to be scattered along the line shown.

The crude net migration statistics hide very much larger gross flows of children. Figure 2.7 shows the proportions of children moving into each part of Britain for every 100 moving out. Thus large numbers of children do move towards areas where fewer are awarded high grades at school, just fewer than those who move out. Even the pattern shown in this figure is a net pattern. If this were the exchange of children between areas every year, many city centres would soon become childless! What prevents that is that these centres

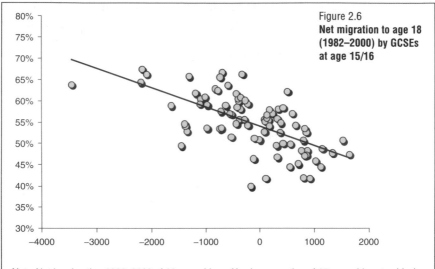

Figure 2.6
Net migration to age 18 (1982–2000) by GCSEs at age 15/16

Note: Net in-migration 1982–2000 of 18 year olds on X axis; proportion of 15 year olds not achieving five A–C GCSEs on Y axis; each point is a 1999 European parliamentary constituency in Britain.
Sources: As for Figures 2.5 and 3.4.

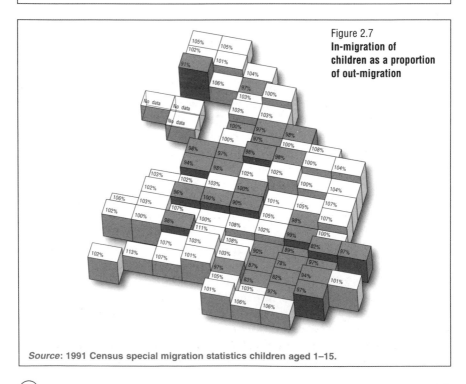

Figure 2.7
In-migration of children as a proportion of out-migration

Source: 1991 Census special migration statistics children aged 1–15.

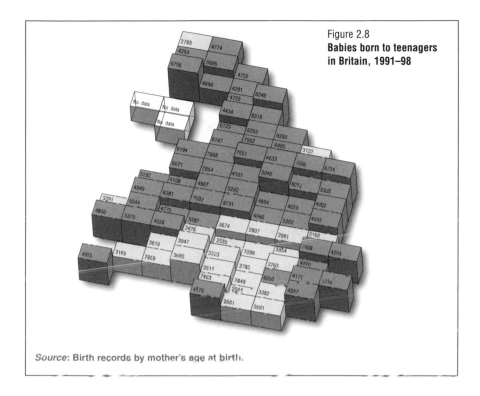

Figure 2.8
Babies born to teenagers in Britain, 1991–98

Source: Birth records by mother's age at birth.

are repopulated by births (Figures 2.8 and 2.10 respectively show how very high numbers of births are added to Central London's population each year, explaining how a 78% replacement rate through in-migration can be maintained over time). It is babies, younger children, the children of immigrants and the children of families who find they have become poorer who replace most of the more affluent children who are leaving the city centres every year. The overall geography of where children live is altering slowly too, but the vast majority of annual migration is required just to keep the social system static – so that affluent areas stay affluent and poor areas remain poor.

Babies born in 1982 don't just grow older and move. They too have children and a few have children much younger than others. There is a very strong geographical pattern to Figure 2.8, which shows the number of babies born to teenage mothers in each area between 1991 and 1998. The map is an approximation of the distribution of 18 year olds in 2000 who are themselves already parents. There is a very clear north–south divide shown though the map, dividing areas where fewer than 4000 babies have been born to teenage mothers from areas where up to over 8000 babies have young parents. The East of London

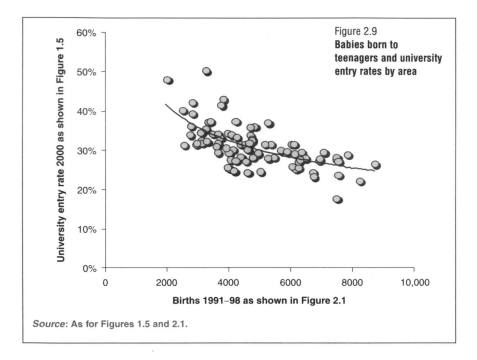

Figure 2.9
Babies born to teenagers and university entry rates by area

Source: As for Figures 1.5 and 2.1.

is the main exception to this divide where rates of teenage pregnancy are comparable to the North of England, Scotland and Wales (we can talk of rates as these areas have roughly equal populations). Compare Figure 2.8 to Figure 1.5 in the previous chapter of the proportion of 18 years olds entering university in 2000. The two maps appear to be rough mirror images of each other. The places from which children are more likely to get to university are the places where they are least likely to have children young. Are there really such different geographically separated groups of children in Britain following markedly divergent paths through life?

Figure 2.9 provides a clearer way of assessing whether the two geographical distributions do tend to be the inverse of each other than can be achieved by visually comparing the two maps on which the figure is based. It suggests that there is a clear inverse relationship with all areas in Britain lying close to that trend. There are no places where a higher than average number of teenagers have children and a higher than average proportion go to university, and no areas where few teenagers have children and a low proportion go to university. This is a little surprising as these are large geographical areas and only a minority of children go to university (in all but one of the places) and only a very

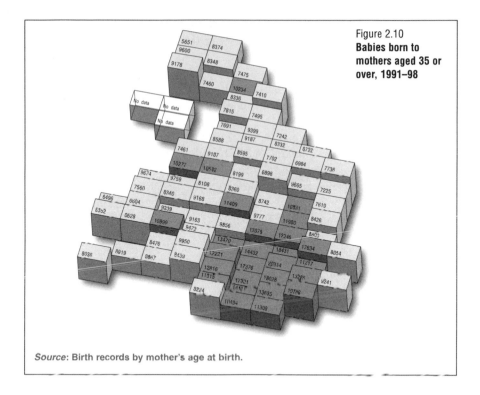

Figure 2.10
Babies born to mothers aged 35 or over, 1991–98

Source: Birth records by mother's age at birth.

small minority of teenagers have children. Most children by the age of 18 neither go to university nor have children of their own (although a very small number do both). The graph suggests that similar processes underlie what makes it more or less likely for children to take different paths through life and these processes are at play across the whole country and are strong enough to result in quite a uniform distribution of life chances by area. Everywhere with more than 6000 births to teenagers has a university entry rate below 30% in 2000. Nowhere where the university entry rate is above 40% have more than 4000 children been born to teenagers over eight years in the 1990s.

The next chapter discusses the extent to which our education system, and what underlies it, is largely responsible for such patterns.

For no group is the last supposition more true than for women who have children later in life. A group who are often portrayed as having the most choices in life often look, in aggregate, to have the least. Figure 2.10 shows how they and their children are crowded into London and the South East of England. Look at

Figure 3.9 to compare this distribution to that of where university graduates end up living in Britain. For many people reading this book, this is the pattern of your mother's geographical choices. For an even larger proportion, this will be where you have children. The more likely you are to be reading this book, the more likely you are to have children later in life yourselves and the more likely are your children to be born in the areas shaded dark grey in Figure 2.10.

A significant number of people have no children of their own during their lives and that number is growing. It is growing partly out of choice, but also as more and more people find it difficult to have children when they first stop using contraceptives to try to conceive at older ages. The area with the lowest proportion of children of all ages between 5 and 17 – and the highest proportion of child bearing age (20–44) – is London Central.

An exercise
(20 to 100 players)

Form yourselves into the map of Britain or your local area according to your place of birth as described at the end of Chapter 1. Now, put up your hands if you moved home in your first year of life. Quite a few of you should have done this. Parents often find the home they are in unsuitable for a small baby, or need to move home urgently to get more space. Now raise your hands if you moved in your second year of life, third year, fourth year, fifth year, and so on. Is there any pattern to when your parents moved home? Did those living closest to city centres or who were first-born children move earlier. What about the direction of travel. How many moved towards the centres of towns and how many into the suburbs? Next, everyone with their hands up needs to move so that the map is reformed by where you had moved to in your first few years.

By now you should all be located according to where you first went to school by age 5. Carry on the exercise, raising your hands if you moved, or moved again at ages 6, 7, 8, 9, 10 or 11. There might be a sudden increase in the later two years. It depends partly on what kind of education system was in place where you lived at this time, but in many areas where you go to secondary school depends on where you live at age 10. Obviously having older brothers and/or sisters complicates that picture, as does families whose parents separate and so on. A more sophisticated form of this exercise would have you playing the part of your oldest sibling. Age 10 is a peak year for migration in Britain. Reform the map to reflect your distribution by age 11.

Carry on to 12, 13, 14, 15, 16, 17. Do you notice any unusual changes in direction here? These are often the ages at which families are most settled geographically, but you might notice quite a few people moving home between age 15 (GCSE year) and the start of A and AS level studies. Finally, if you are in an old university, note the high migration rates of almost all of you at age 18 or 19. For those who doubt the importance of education on migration (and migration on education) ask yourselves why Britain runs a system, unlike almost any other in the world, in which between a third and a half of its 18 year olds are expected to not only leave home, but leave the village, town or city they grew up in and be scattered across the country in a way which appears random, but which you will know (if you have applied to university) is incredibly closely controlled by A level grades and quotas. If you are reading this book in Scotland or Ireland, your experience may be more like that found elsewhere in the world and you are unlikely to have moved as far from home. The same will be true if you are attending a 'new' (i.e. post-1992) university.

Finally go forward in time and predict where you will be at ages 20, 21, 22, 23, 24, 30, 40, 50, 60 and 70. Will you be part of the majority who fit the stereotypes being described here or do you think you may be a little different? (Most people think they will be different. Most of course, are not.)

CONCLUSION

The tour maps in this chapter each show a different relationship between one aspect of life chances in Britain and the underlying landscape of opportunity to attend university. Figure 2.5 shows that there is a general tendency for children to flow downhill between birth and 18 towards areas where they are more likely to win university places. There are exceptions to that rule. It does not hold for the Oxford, Surrey and North West London valley, which was described as the most fertile area of the landscape at the end of the last chapter. As the next chapter makes clear, doing well here often comes at a price that your parents have to pay (Figure 3.6) and many of the children whose parents cannot pay that price fair particularly badly in this part of the UK (Figure 3.3). Thus many parents move themselves and their children out of the valley to 'better schools'. Similarly, there are a few areas, almost all in the south, into which more children are moving than leaving, this appears not to be in aggregate to their collective benefit (if entry to university is seen as important). However, read a little further on and you'll find that these are often good places to be at age 15 rather than 17. The variation shown in Figure 2.6 is largely explicable when such other geographies are also considered.

Figure 2.8, of where teenagers have the most babies, follows the contours of the landscape more closely but also reveals a north–south divide which is stronger than our underlying geography of opportunity. Most teenagers who are likely to become young parents do not go to university whether they have children or not, and so there are other factors at play in creating this pattern of chances. The importance that the landscape retains is revealed by Figure 2.9, but the other factors which account for the variation in that figure are again clearly geographically determined. If they were not, then the spatial pattern shown in the map would not be so clear. The same type of geographically well ordered process has also to be at play for those people becoming parents at the other end of the age scale. These people are geographically concentrated roughly within the landscape of opportunity drawn here. Older mothers fill the valley of greatest university opportunities despite being twice as old at least as most of those going to university. Thus it is not that precise opportunity which leads to them having their children at these ages in these places, nor is it for the chances of their children (they have many years to move home should they wish to influence those by area). A landscape drawn according to a chance which has nothing directly to do with another group of people can still appear closely related to that group. That is because the variable used in this book to draw the human geography landscape is itself a reflection of many other closely related patterns, which are in turn close reflections of other facets of society. Where children are most encouraged and cajoled to study hard, where their schools are most likely to be good, where their teachers have more time and are perhaps a little better equipped, where their parents are most able to help them and have the greatest incentives to do so, where children are most expected to go to university, is where most do. And what might underlie all these things? Most probably the same forces that encourage women in these places to have children late in life. The next seven chapters explore just a few of the forces, facets and relationships revealed by the human geography of the UK.

3 Education

...the sorting out of children

In most schools in Britain children are sorted into groups according to how able they are deemed to be. They are tested at various points through examinations and those test results are used to sort them further and to determine whether they will be permitted to progress on to each next stage after the end of compulsory schooling at age 15. For those who leave school at age 15, 16, 17, 18 or 19, the results of these tests are used to curtail their employment or further education options. Then, for the minority who go to university from age 18, they are subjected to further tests. However, university tests are designed so that almost no one is allowed to fail, but students are still crudely graded at age 21 (more than 90% are awarded what is called a second-class degree). Thus, by age 21, almost every child in Britain has been labelled as inadequate in some way – from not being awarded any qualifications at all, to being labelled as second class. This might appear a little odd, especially if you are not familiar with the way things are done in Britain. The overall effect is to put almost everybody in their place socially through a system labelled education. What then determined what place you are put in?

Throughout this book I have been assuming that you are aged between about 16 and 20 and are reading this in Britain in the first decade of the twenty-first century. Given that you are reading this book, you almost certainly took General Certificate of Secondary Education examinations (GCSEs) when you were aged around 15/16. If you are reading this book from outside Britain, it might be useful to know that these are also known as Key Stage 4 exams. A National Curriculum came into force in Britain in 1992 to ensure uniformity in what all state school pupils were taught and by 1995 pupils aged 11 at primary schools were beginning to be examined nationally at what is called Key Stage 2. In many ways Key Stage 2 represented the reintroduction of the old 11 plus examination which determined many older people's chances in life (and which still exists in a few parts of the country). Key Stage 2 differs in that

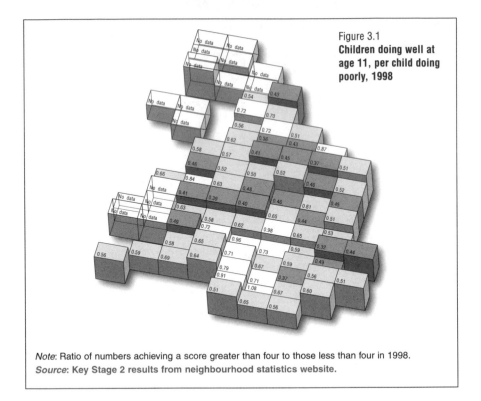

Figure 3.1
Children doing well at age 11, per child doing poorly, 1998

Note: Ratio of numbers achieving a score greater than four to those less than four in 1998.
Source: **Key Stage 2 results from neighbourhood statistics website.**

it is not a 'last chance' as the old 11 plus became, but the first important exam of a series of exams for what have probably become the most examined set of children in the world. The first of this 'most examined generation' were the children who took KS1 in 1991, KS2 in 1995, KS3 in 1998, KS4 in 2000, AS levels in 2001, A Levels in 2002, and – for the minority who had got this far – entered university in summer 2002. They were the first-year university students I was teaching while I wrote this book. They looked a little washed out and were a little better behaved than their predecessors were, but they were no more clever or imaginative or insightful than them. What then was the purpose of so many examinations? To see this we need to look at where children are failed at each key Key Stage of the process.

For most children in Britain the first examinations that really matter are those taken at age 11. They influence the 'sets' they will be placed in later in school and if you are not placed in a high set it is very hard to do well later on. You might have thought that at this age we would be encouraging children, but not so. Figure 3.1 shows the number of children deemed to have done well in

HUMAN GEOGRAPHY OF THE UK

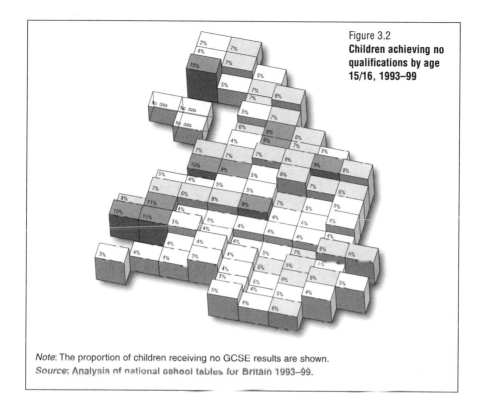

Figure 3.2
Children achieving no qualifications by age 15/16, 1993–99

Note: The proportion of children receiving no GCSE results are shown.
Source: Analysis of national school tables for Britain 1993–99.

these exams compared to those deemed to have done poorly (it excludes those who gained an average grade of 'four'). Only in one constituency (Surrey) are slightly more children awarded grades to indicate that they did well as opposed to poorly. The other extreme is found only a few miles away across the capital, where three children are deemed to have done poorly for every one who does well in London North East. At lower geographical levels the extremes are closer still. Parliamentary constituency ratios are highest in England in Sheffield Hallam, where 2.2 children do well for every one who does badly, and lowest a few hundred metres away in Sheffield Brightside at 0.1. The Sheffield average, of 0.45 means little for the most educationally divided city in England. At a lower level still, half of all the children doing well in Sheffield city lived in just seven of the city's 29 wards in 1998. Scotland and Wales are blank on the map because, probably sensibly, their parliament and assembly do not favour releasing this information.

Around 200,000 children left school with no qualification from the areas shown in Figure 3.2 between 1993 and 1999. This figure can largely be predicted from the proportions deemed to have failed Key Stage 2 some five years earlier. However, there are nine constituencies where such an exercise would under-predict the proportion (by between 3.7 and 2%: Merseyside West, Leeds, Greater Manchester Central, Yorkshire West, Tyne and Wear, Yorkshire South, Sheffield, Cleveland & Richmond, East Yorkshire & North Lincolnshire) and three where the map would be over-predicted (by 2.3 to 2%: Cornwall & West Plymouth, Suffolk & South West Norfolk, Essex North & Suffolk South). It is mostly the impact of migration which helps to make Figure 3.2 a map with a simpler pattern as compared to that seen in Figure 3.1. The very high proportions of children who left school with no qualifications in the Welsh valleys and Glasgow perhaps explain a little more about why educational statistics there are a little harder to come by. How is it possible that so many children were leaving school without a single qualification while the numbers receiving university degrees were accelerating (Figure 2.3)? The data used here does not include children attending those special schools where GCSEs are often not taken. Almost all these children were capable of passing some exams, but none passed.

A further 200,000 children left school with only a few qualifications, less than five GCSEs at any level. In Figure 3.3 they are combined with those receiving no qualifications, in total over one in seven of all children in Britain. Such low qualifications will disbar these children from most jobs that require developing further skills such as secretarial work or working for a large firm in the building trades. Figure 3.3 is not a map of ability. Children in Glasgow are not many times less able than those growing up in the Highlands of Scotland, as the map might imply. It is instead a map of migration and money. In the Highlands of Scotland, due to the sparsely populated nature of that land, parents have little choice where to send their children to school and so there are no schools deemed to be 'failing'. In a compact city such as Glasgow, short geographical distances and often money allow parents to send their children to particular schools, allowing other schools to 'fail' disastrously. Parents can migrate out of the city while still commuting to work within it and so many take their children away from areas thought to be detrimental to their education. Underlying all of this, however, is the nature of these examinations. These examinations are designed to fail a certain proportion of children and to give low grades to yet more.

Examinations where almost no one fails are possible and are the norm where students pay for their education in universities and private schools. People will not pay to be given a label saying they have failed. However, if examinations did not label a proportion of children as failing, then it would not be

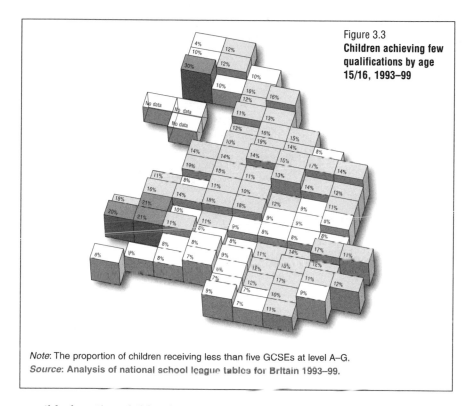

Figure 3.3
Children achieving few qualifications by age 15/16, 1993–99

Note: The proportion of children receiving less than five GCSEs at level A–G.
Source: Analysis of national school league tables for Britain 1993–99.

possible for other children's parents to claim that they had 'done well'. Why we fail quite so many is more difficult to understand. Figure 3.4 adds to Figure 3.3 all those who did not achieve what was seen as the basic school-leaving qualification of at least five GCSEs at levels A–C. This is loosely equivalent to the old school-leaving examination that existed before O levels were introduced. More than 55% of all children failed to reach this level in the period shown here (to include Scotland, which uses a different examination system, it is assumed that a standard grade 2 is equivalent to C and a 4 to G). The proportion in Britain failing at this level has fallen from 59% to 51% over the period considered here. This apparent achievement is illusionary because, at the other end of the education ladder, more are going on to take degrees, Masters-level qualifications and so on. As the bottom rungs of the ladder are lowered, the top is stretched further and further into the distance. In the near future achieving only five GCSE A–C grades will be seen as failure.

For some children in Britain achieving only five GCSE grades at A–C has already been viewed as failure. Most obviously these included the 7% of

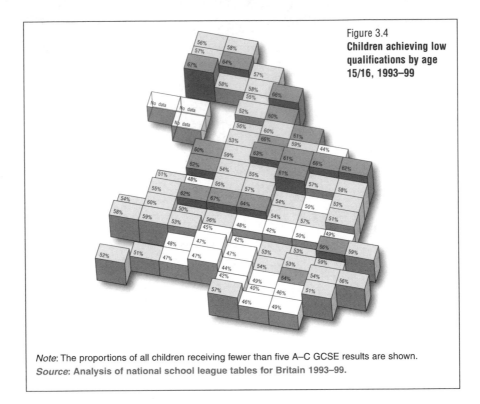

Figure 3.4
Children achieving low qualifications by age 15/16, 1993–99

Note: The proportions of all children receiving fewer than five A–C GCSE results are shown.
Source: Analysis of national school league tables for Britain 1993–99.

children who attended private schools between 1993 and 1999 (many more attend now). Figure 3.5 shows that 83% of private school pupils achieved at least this level, more than twice the proportion attending comprehensive schools and three times those sent to secondary moderns (schools which are a relic of the old 11 plus system). Although much is written on parents who opt out of the state system to send their children to private schools and the supposedly unfair advantage they then gain, their numbers are dwarfed by the quarter of all children who took their GCSEs in a state-funded school that practised some kind of discrimination on entry. Many of these are the so-called church schools. In general these schools exclude what they see as less able children, and consequently a majority of their pupils in the 1990s did well at GCSE by the standards of those days. They also had the lowest proportion receiving no qualifications of all the four types of school. Today many more children attend discriminatory (or selective) state-funded schools, and many more attend private schools than did in the 1990s. For those who doubt the effect of selection, whether it is state-funded or private, the figures for secondary modern schools show what occurs when a school is truly

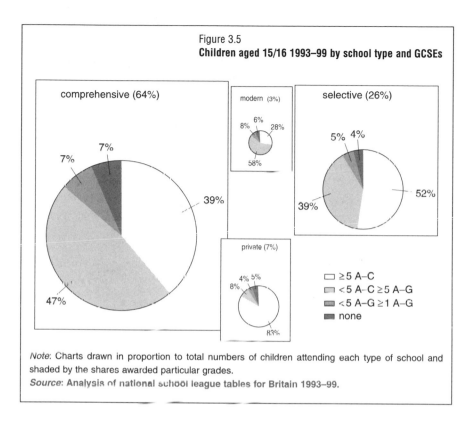

Figure 3.5
Children aged 15/16 1993–99 by school type and GCSEs

comprehensive (64%)

modern (3%)

selective (26%)

private (7%)

≥5 A–C
<5 A–C ≥5 A–G
<5 A–G ≥1 A–G
none

Note: Charts drawn in proportion to total numbers of children attending each type of school and shaded by the shares awarded particular grades.
Source: Analysis of national school league tables for Britain 1993–99.

labelled as being a 'failure'. Almost three-quarters subsequently 'fail'. British education has been set up in such a way that children can only succeed if others are deemed to have failed. Success can be aided by migrating, and millions of children are moved to new homes every year to aid this process. More crudely, allowing schools to discriminate with their entry can aid success and success can also be paid for through the private system.

Although only 7% of children attended private schools at age 15 in the 1990s, that proportion disguises highs of 22% in London Central and lows of near 0% in several other places (Figure 3.6). Locally the proportions attending private schools are highest where the failure rate is highest in Figure 3.3! Parents fear for their children's life chances most in these areas and so opt to try to pay to improve them the most there, further damaging already disadvantaged local state schools by withdrawing their children. However, private education is limited to those able to afford to pay. The proportion of children attending

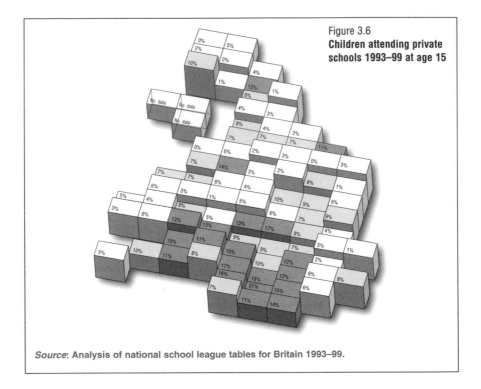

Figure 3.6
Children attending private schools 1993–99 at age 15

Source: Analysis of national school league tables for Britain 1993–99.

private schools in Edinburgh is higher than that in Glasgow, as there are more wealthy people in the Scottish capital. Across Britain, private schools dominate in the south of England, even in areas where results overall appear fine. The pattern of high private school attendance is uncannily similar to that of GCSE success, the inverse of Figure 3.4, which, in turn, is worth (bearing in mind Figure 2.4) comparing to Figure 1.5 (those who go to university). Money can largely buy your children out of the secondary school education lottery, but what exactly is it buying them into? Do children from private schools who attend university feel that they have succeeded or simply that they have just achieved what was expected of them? If so, what then is the next hurdle that we place in front of them in the name of education?

Before turning to further and higher education it is worth briefly considering how current trends at age 15/16 might alter the future map of so-called educational achievement in Britain. Figure 3.7 shows the annual fall in the proportion of children failing to achieve the basic standard of five A–C grades between 1993 and 1999. Each year a slightly higher proportion of children are

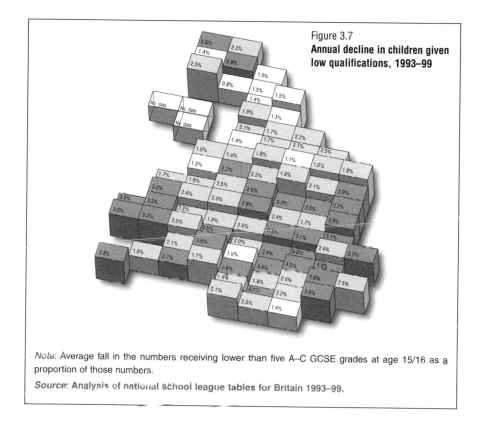

Figure 3.7
Annual decline in children given low qualifications, 1993–99

Note: Average fall in the numbers receiving lower than five A–C GCSE grades at age 15/16 as a proportion of those numbers.

Source: Analysis of national school league tables for Britain 1993–99.

allowed to pass exams than the year before. Whether this reflects improving ability, better teaching or lowering standards is unimportant to children's life chances as the worth of these exams falls in almost direct proportion to the additional numbers allowed to pass them. However, this 'improvement' is not evenly spread across the country when expressed as a decline in the proportion who 'fail'. It is highest in the south of England and Wales. It is lowest in the north and in most of Scotland. It is low, too, in odd places such as Hampshire North & Oxford (but here rates were good to begin with). This figure has to be read in conjunction with Figure 3.4 to see both where children's life chances, as influenced by GCSEs, are best and worse, and whether they are improving or worsening in relative terms. The changes shown in Figure 3.7 will influence the lives of millions of adults in Britain for decades to come.

For those who are deemed to do well at 15, what and where next? The numbers of students in sixth-form colleges, further education institutions and

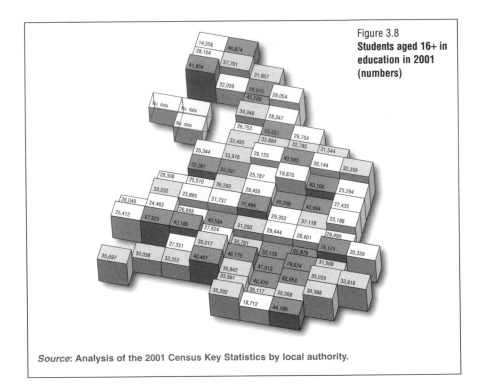

Figure 3.8
Students aged 16+ in education in 2001 (numbers)

Source: Analysis of the 2001 Census Key Statistics by local authority.

higher education institutions are shown in Figure 3.8 – some three million people in 2001. The geographical clustering is due to the locations of large universities. Half a million people aged 16 or over are students in London. There are both more universities and a higher proportion of children staying on at school in the south and so the numbers are generally higher there. But have they succeeded? Of those who stay on at school but who do not go to university, most will be awarded relatively low grades, implying failure as compared to their peers. They may be qualified to work behind the counter in a bank or join the police force. Very few children who gain high A level scores leave school on a winning streak and find a job. Instead, almost all now go to university. Once there they at least know they are almost certain to be awarded a degree if they do not drop out, and this will qualify them for jobs their counterparts cannot dream of applying for (fast-track promotion in a bank or the police force, for instance), although many will at least initially end up with menial jobs such as working in a call centre (especially when the labour market is adjusting to the increase in graduates). Almost all will be awarded what's called a second-class degree, worth far less than that of their parents if they too went to university.

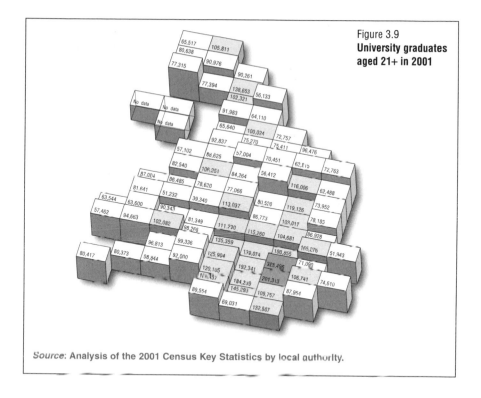

Figure 3.9
**University graduates
aged 21+ in 2001**

Source: Analysis of the 2001 Census Key Statistics by local authority.

Where next for the minority of children who are awarded a degree? A lot become doctors, specialist nurses, teachers, managers, engineers and the like. Occupations often deemed more useful than cleaners, train drivers, builders and shop assistants (although how society would operate without the latter is hard to imagine). However, Figure 3.9 shows that the largest geographical concentration of university graduates, almost a third of a million, are living in central London. Most of these people are engaged in making money out of other people: merchant bankers, accountants, international financiers, estate agents, staff of private firms' head offices and so on. Half of all graduates in Britain live in 30 European constituencies, only three of which are not in southern England. The meticulous sorting out of children results in eventually selecting this minority, many originally from the south of England, and enticing them to move to (or back to) particular parts of the south. The highest numbers of all are found in and around London; other areas have three to six times fewer. Places cannot run without graduates, but they can get by with 50,000, as the map shows (roughly a tenth of their populations). Ironically, perhaps, the greatest concentration of university graduates is in the same

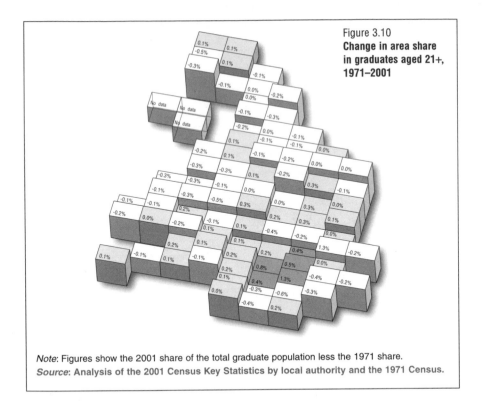

Figure 3.10
Change in area share in graduates aged 21+, 1971–2001

Note: Figures show the 2001 share of the total graduate population less the 1971 share.
Source: Analysis of the 2001 Census Key Statistics by local authority and the 1971 Census.

small area of the country where 8% of children receive no qualifications, 53% did not attain five GCSEs A–C and 22% of children are sent to private schools by their parents. A majority of parents having a university degree does not privilege a place educationally for everyone living there.

Lastly in this series of figures, how is the map of graduates changing? Figure 3.10 shows this. There were more than three times fewer graduates in 1971 as compared to 2001, but their geographical distribution was not dissimilar. In 1971 3.5% of all graduates lived in London Central, by 2001 that had risen to 4%. Figure 3.10 shows that changing distribution is further concentrating graduates into London to the detriment of the north, Scotland, Wales and the coastal fringes. Thirty years of raising education 'standards', encouraging huge numbers of children to stay on at school (including raising the minimum age at which you can leave from 14/15 to 15/16), opening the university doors wider and giving out very many more degrees has resulted in a slight concentration of the pattern seen before all this began. Our education system has been changed since 1971 in such a way as to maintain a process that sorts children into groups and then encourages those groups to move to particular places, mostly the same places as before, though now a little more where there were most to start with. If so many more people are apparently learning so much more than people did in the past and passing so many more exams, why have the geographies changed so little over time?

An exercise

(20 to 100 players)

Form yourselves into groups, each group containing just a few (three to six) students. You could use the map you made at the ends of Chapters 1 and 2 to form the groups so that they are made up of students who come from near one another, but that is not what the group should have in common. Instead, each group needs to imagine that all their members are aged 30, have children of their own about to enter the education system and have somehow come to political power. As a group, you have ten minutes to complete the following task before you will have to present your arguments to a neighbouring group: 'Design an education system where the aim is to teach children rather than sort them. What role would exams take in such a system, if any? At what ages would you examine each child's ability, on what subjects/issues and what proportions would you decide to fail at any stage? How would you decide who goes to university and which university they go to? How would you then grade university students?'

After ten minutes present your arguments as a group to a neighbouring group in the room. They, in turn, should present their suggestions to your group. Vote on the result and carry your combined most popular opinions forward to repeat the process after a further five-minute discussion as a combined larger group of students. Then combine groups again and again until one set of ideas has won out. What led the most popular system to win through? Was it a good system, or simply well presented? Did you design it assuming your children were 'able' (most parents think their children are above average!)?

Finally, take your initial groups and, at random, assign each group an ability level. This is the level your prospective children could be expected to achieve under the current education system. One-third of all groups are now made up of the prospective parents of children who would attend university under the present system, one-third will not gain the qualifications to attend university but will be awarded five GCSE A–C grades, and the final third are the prospective parents of children who will not achieve this under the current system. Now, with your imaginary children in mind (and their interests at heart), each group needs to decide which of the systems that were presented, it now thinks is best. Each group gets an equal vote. Vote on each system. Which system now wins?

CONCLUSION

Surely some children 'succeed', you must be thinking, as not all are failed by education in Britain. Well the above account is perhaps a little unfair on all the efforts of millions of school children, teachers, examiners, the writers of textbooks, and designers of education websites and television programmes to educate. My view is that, largely despite the way in which schools are organised and examinations imposed, many children still manage to learn, many teachers still manage to teach and far more people are better informed about the world than they were a generation ago. However, the basic purpose of the education system is still to put children in their place. The key selection point has moved from age 11 in the 1960s to age 17 now. University admissions officers and those who control them are the gatekeepers to middle-class entry. A generation ago boys from Britain's top public schools did not need to go to university to maintain their status. Now even they have to go and it would be almost unthinkable that a member of the royal family did not attend university. However, it would also be unthinkable if they were only allowed entry to the 'wrong' kind of university and so we still run an ancient military academy to take those who need to avoid such a fate!

Entry to university by A levels is simply entry largely according to school, which reflects parents' wealth through their ability to pay for their children's cramming for exams (by directly paying their school, through extra private tuition, or by paying more for their house in a better catchment area). Whenever someone suggests altering university entry systems there is uproar because the stakes are so high (the privileges of the children of the privileged). Even if access is widened and more children from poorer backgrounds are allowed to attend university, a new divide is forming at age 21. Increasing numbers of students take a Master's degree after their first degree. Almost all of these have to be paid for privately. One, at least beneficial, effect of having an almost entirely private tier of education at age 21 is that it is at this level for the first time that a majority are not labelled as having failed in some way.

Once you pay for your piece of paper, almost all of you are either passed or receive a distinction. There are no more second-class degrees when such large amounts of money are involved. The key age of selection and median school-leaving age increases by about a month every year. It currently stands at around age 17. Half way through this century it should reach age 21 if the past century is any guide. The selective university sector is ever so slowly transforming itself away from the training ground of the elite and into what will become compulsory tertiary education – if current trends continue. Tomorrow's children face yet more years of examinations and an even more drawn-out process of being sorted out through education. That sorting out may well only be finished in the near future in their early twenties through postgraduate qualifications. And, as now, it will not be by their efforts that they are sorted but by their access to the resources needed to achieve such qualifications – their parents paying for those postgraduate degrees.

4 Identity
...labelling people and places

People are given many kinds of label, and label themselves and others in many ways for many purposes. Some labels are harder than others to shake off or disguise – your sex and age, for instance. Others it is easier to keep confidential and hide if you wish, such as your religious beliefs and sexuality, although people may well make assumptions about you. Areas too can be labelled based partly on the labels assigned to the people who live, have lived or will live within them. 'Gentrifying', 'growth', or 'up-and-coming areas' are areas labelled through an assumption as to who their future inhabitants will be. Labels are central to studying the human geography of Britain, in both trying to understand how labels come to stick and what useful meanings they might have. There is often a great deal of truth behind what is sometimes called stereotyping areas. There is also, of course, a great deal about the population of an area and the people with whom they interact that cannot be captured by a label, just as labelling individuals simplifies and masks their lives to a huge extent. However, a large part of the business of running society involves labelling. Chapter 3 could have been entitled 'the labelling of children'. Most labelling may well be unfair and unfortunate, but it is through labelling that society is run, that tasks are allocated (usually unfairly) and that the rewards of our efforts are distributed (almost always unfairly). Labels matter. They may all be ephemeral but that makes them even more crucial to understand. Two centuries ago most social statistics did not include sex; it was irrelevant because women were seen not to matter. In two centuries' time they might not include sex again because sex will not matter. We can begin to trace, through the labels that are currently deemed to be of most importance, those categorisations which are used to control, blame and encourage people, and, above all, which are used to attempt to ensure that they conform. In this chapter, ten key labels, which are taken from the 26 'Key Statistics' tables first released by the government from the 2001 Census, are examined. Almost all of the other 16 tables are used elsewhere in this book (Figures 3.8–3.10 have already used this same source). Government does not choose themes at random to publish. It and the quasi-official bodies it consulted are convinced

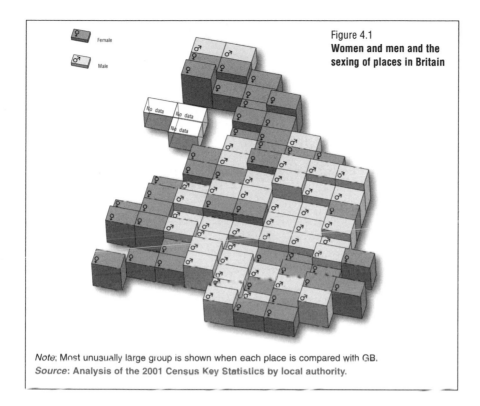

Figure 4.1
Women and men and the sexing of places in Britain

Note: Most unusually large group is shown when each place is compared with GB.
Source: Analysis of the 2001 Census Key Statistics by local authority.

that these are key things which should be counted and published. Other statistics are counted and not published, such as those on income and wealth. In this chapter, however, are labels that the powers that be wish us and our neighbours to know about. The ten labels range from what you might just think of as simple facts about yourself, to categories that you might object to having been put in. They are based on your occupation (or lack of one); the colour of your skin; your thoughts about a god; where you have come from and how alien that place and you appear to be; your living arrangements; who you share your bed and home with (or used to share with); and how posh or common you are deemed to be. None of this information was collected clandestinely. It was all derived from the Census form. Admittedly, the person in your household who completed that form could face a £1000 fine if they did not declare that information. Few people, however, may realise just how effective one little form duplicated almost 30 million times can be.

Figure 4.1 uses the simplest census question to begin this illustration of place labelling. On the Census everyone in Britain is labelled as either female

(29,345,507 people) or male (27,758,420 people). The proportions are not equal because more men are born (Figure 2.2) but more men die young than women and differing numbers leave and enter the country at different times (only a tiny number change sex). Nationally, by 2001, there were 94.6 men for every 100 women. The figure shows area deviations from that national proportion. The 41 areas with more than 94.6 men per 100 women are labelled ♂ and the 43 areas with more women are labelled ♀. This is a very crude form of area labelling. In theory, the presence or absence of a single man or woman in any area could swap its label over. However, in practice the method is quite robust. This is an unambiguous categorisation that gives a sex to all places. Only 15 places, categorised in this way, changed their sex over the last decade and there is a pattern to the sex changes. Five mainly rural places became unusually male as women left the place, and these women turned ten largely urban places unusually female. The most female place (52.9% female) in Britain today is Glasgow, the most male place is Thames Valley (50.1% female). At this scale, the gap of 2.8% is tiny, but with sex small things matter greatly.

Just as places can be labelled according to the sex of their inhabitants, so too can they be labelled by their ages. Figure 4.2 uses the age categories included in the Key Statistics (except for amalgamating ages over 75 to allow comparison to be made with the past Census figures). A place is given a specific age label if that age group is more in excess of its national proportion in that place than is any other age group there. Thus London East is labelled '0–4' because that age group made up 6.5% of the population of that area in 2001 compared to 5.9% of the national population. The gap of 0.6% is greater than for any other age group in London East. To be the atypical group that is used as an area label it helps if the group is both large in number and unevenly spread across the country. It is difficult for one particular group to be both things and so a diverse set of labels appear in these figures. Ten years ago London East was labelled '65–74', as was Yorkshire West (another of the three baby areas) in 1971 and Birmingham West was much older a decade ago too. The labels of these areas have changed over time. Most areas were categorised by the same age group ten years ago (so the place remains young or old as the people enter and leave it). Of those that have changed, most have aged slightly as the population as a whole ages and the local population has not been replaced by younger in-migrants.

Ethnicity is a label that was first used in the Census in 1991; it is in use in Britain as a euphemism for race, which, for the Census authorities in Britain, is seen as skin colour. Why the question came to be asked first in 1991 is a long and interesting story, but here we are most interested in the effects of that labelling. Figure 4.3 uses the 2001 categories included in the Key Statistics

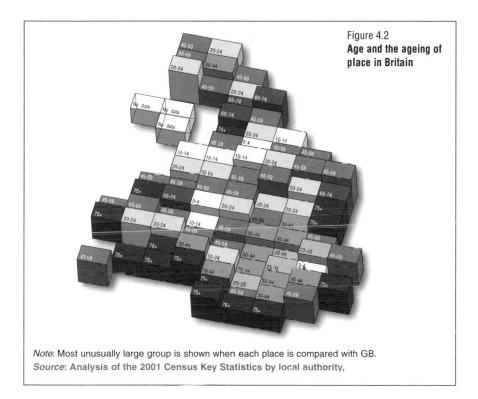

Figure 4.2
Age and the ageing of place in Britain

Note: Most unusually large group is shown when each place is compared with GB.
Source: Analysis of the 2001 Census Key Statistics by local authority.

amalgamated to their 1991 equivalents to allow the changing ethnicity of place to be assessed through the effects of these labels (the 2001 Census included 'mixed' categories as the labels were changed to incorporate lighter coloured skins, including the Irish). Four of the ten comparable ethnic groups included are too evenly spread across Britain to label even a single place. One group, 'White', is so unevenly distributed that some 57 of the 84 areas have the highest excess as White. Ten years ago it was exactly the same 57 places that were 'White'. The only changes to place labels have occurred within the minority of places labelled after 'minorities'. London Central, London South Inner and London North East were Black Caribbean in 1991; London North and London South East were Indian. These five areas' new colours can be read off the Figure. That three times as many places changed their sex over the same period is telling. Ethnicity is a label that sticks geographically.

Just as the Census authorities introduced ethnicity as a new form of official identity in British social statistics in 1991, they – or more strictly speaking a few influential members of parliament – introduced religion into the 2001

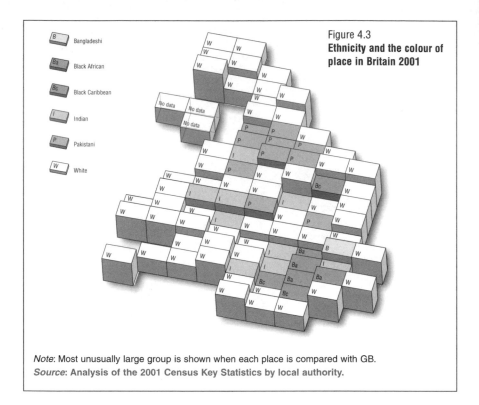

Figure 4.3
Ethnicity and the colour of place in Britain 2001

Legend:
- B — Bangladeshi
- Ba — Black African
- Bc — Black Caribbean
- I — Indian
- P — Pakistani
- W — White

Note: Most unusually large group is shown when each place is compared with GB.
Source: Analysis of the 2001 Census Key Statistics by local authority.

Census. Strictly speaking (again), it had been asked before, but only in a special Census held some 150 years ago. Again the story behind this label is intriguing, but the motives of those involved are best revealed through the results. Three of the nine possible religious labels belong to people in groups either too small and/or too evenly spread across the country to label places (Buddhists, people of 'other religions', and people who left the question blank – which was by law permitted only for this question). Of the six religions left, 'Christian' is the most unevenly spread, labelling a narrow majority of 43 areas, while the next largest group, of 22 areas, contain unusual numbers of people explicitly saying they are of no religion, or were Jedi, who were shamefully amalgamated with the atheists by the Whitehall unbelievers. Geographically, 'no religion' matches the old (Wales and Scotland) and new (university city) non-conformist areas. Of the remaining areas in Figure 4.4, nine are Muslim, five Sikh, three Hindu and two Jewish. Within these 19 areas the beliefs of only 8% of their populations are used here to label some 13.4 million people's areas. Is this reasonable?

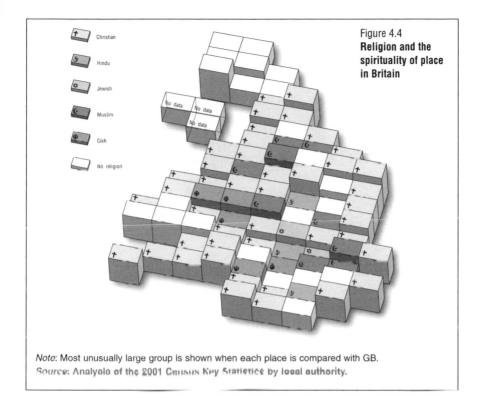

Figure 4.4
Religion and the spirituality of place in Britain

Christian
Hindu
Jewish
Muslim
Sikh
No religion

Note: Most unusually large group is shown when each place is compared with GB.
Source: Analysis of the 2001 Census Key Statistics by local authority.

Sex, age, ethnicity, religion, what next? Are you married and just looking for fun or single and more serious about trying to start a relationship? If you are wondering what I am talking about, have a look at how people describe themselves in lonely hearts columns. Identity is not just an official construct, we construct and publicise our own and others' identities daily. Identities also intermingle. Ethnicity, religion, age and sex are not independent. An old person is more likely to be female, White and Christian in Britain than is the population as a whole. She is also very much more likely to be widowed than is everyone else. Figure 4.5 shows the unusual in the geography of our quasi-legal/ religious state of sexual availability and nest building (and breaking) by area. People, it would appear, are much more ready to remarry in the south of England as they age and move out of both cities and their initial relationships. The unusually high numbers of divorced people in three northern areas is not a reflection of the divorce rate but rather the reluctance of divorcees there to remarry. In the north too, in the valleys of North Wales and Northumbria are the unusual numbers of widows (and a few widowers). There

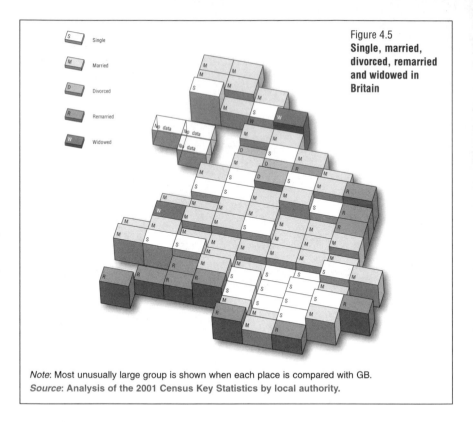

Figure 4.5
Single, married, divorced, remarried and widowed in Britain

Note: Most unusually large group is shown when each place is compared with GB.
Source: Analysis of the 2001 Census Key Statistics by local authority.

were, of course, once working slate and coal mines here too and other industries that contributed to very high premature male mortality in the past.

There is an industry called geo-demographics which classifies people according to their postcodes. Figure 4.6 uses much larger areas than these and shows the eleven groups from within a classification of households that have a geography distinct enough for a place to be labelled after them. They have been ordered in crude life-stage order. To illustrate how this works, pretend you are the extremely atypical person who behaves exactly socially and geographically as the geo-demographers would wish. You could, for instance, be born to a lone parent [A] in Greater Manchester (East) who then moves in with her mum [B] in Yorkshire West before she finds a man, marries [C] and settles down back in Lancashire South. Not wanting to go too far from home you become a student [D] in Sheffield, get a job and live in a mixed household [E] in London North before saving enough for the deposit on your own flat [F] slightly further out in London North West. You meet your soul-mate, sell

HUMAN GEOGRAPHY OF THE UK

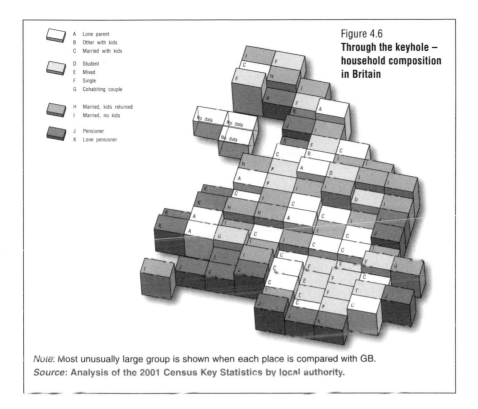

A	Lone parent	
B	Other with kids	
C	Married with kids	
D	Student	
E	Mixed	
F	Single	
G	Cohabiting couple	
H	Married, kids returned	
I	Married, no kids	
J	Pensioner	
K	Lone pensioner	

Figure 4.6
**Through the keyhole –
household composition
in Britain**

Note: Most unusually large group is shown when each place is compared with GB.
Source: Analysis of the 2001 Census Key Statistics by local authority.

the flat and move into their place [G] in Bristol. I'll let you complete the story, save to say you'll end up alone on the coast. Again, of course, it may be very small but unusual extra numbers of households living in a place that lead it to be labelled in this way. Wales is not a land of lone pensioners and single parents as Figure 4.6 might suggest! These are just the two groups most over-represented in the five Welsh European constituencies. Geo-demographics emphasises what is unusual about a place. In truth, there are more similarities than differences between places.

Did you notice all that moving about your hypothetical geo-typical alter-reality engaged in? That was not extreme. For places to have social definition people need to keep moving, as Chapter 2 explored in detail for children. Figure 4.7 shows one version of the within-Britain migration picture for people of all ages, using the same technique that has been used throughout this chapter of labelling places according to the group from one of the Census Key Statistics tables which is most in excess in the area compared to its numbers nationally.

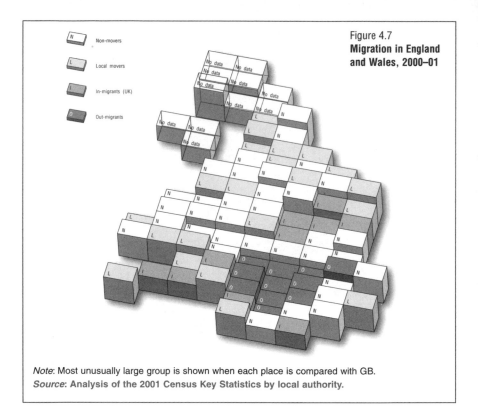

Figure 4.7
Migration in England and Wales, 2000–01

Non-movers

Local movers

In-migrants (UK)

Out-migrants

Note: Most unusually large group is shown when each place is compared with GB.
Source: Analysis of the 2001 Census Key Statistics by local authority.

However, the groups used are not mutually exclusive and exhaustive. People can be (and are) both in-migrants and out-migrants between areas at the same time. Data has yet to be published for Scotland as I write. In England and Wales the most common label is that a place is typified by people who don't tend to move house. These are the suburbs and, just a little further out of town, the ex-urbs where families settle for years. Next most common are places where people tend only to move locally if they do move, swapping houses for an extra bedroom or a 'better area'. Almost equal are the last and smallest two categories: areas where an unusual number move in from outside the local district (these form a ring around London) and the areas they move from (London), often at the cost of their time in much more commuting.

What, though, of people who move into and out of Britain? The accuracy of the 2001 Census is largely based on a quite well-informed guess that at least a million more people left these shores in the late 1990s than we first thought had (before the counting of who was here began in 2001). For those coming

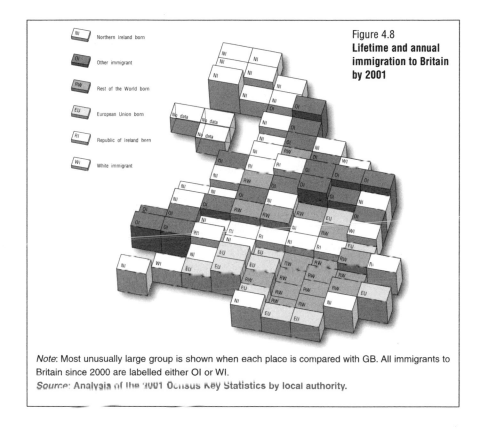

Figure 4.8
Lifetime and annual immigration to Britain by 2001

Legend:
NI — Northern Ireland born
OI — Other immigrant
RW — Rest of the World born
EU — European Union born
RI — Republic of Ireland born
WI — White immigrant

Note: Most unusually large group is shown when each place is compared with GB. All immigrants to Britain since 2000 are labelled either OI or WI.

Source: Analysis of the 2001 Census Key Statistics by local authority.

in we have a much better idea of their origins and locations. Figure 4.8 stretches the method used so far in this chapter even further by combining overlapping statistics from two tables: one of where people were born (if overseas) and their ethnicity (if recent arrivals). The figure can be characterised as describing a series of rings of immigration past and present that radiate out from London. London is typified as being wholly made up, and largely housing, unusually high numbers (for Britain) of people born outside the European Union (EU). Surrounding it is a ring of areas of people born in the EU (excluding Britain and Ireland). Surrounding that ring are areas to which Irish-born immigrants have moved since settling in Britain many years ago. Above them in 2001 was a northern and Welsh band of (resettled from London) non-white recent arrivals, and above that Northern Ireland emigrants dominate. The 'Other immigrant' belt is partly of where people who had recently arrived in this country, often fleeing persecution in poor countries or wars (many of which we were partly involved in), were settled by the authorities.

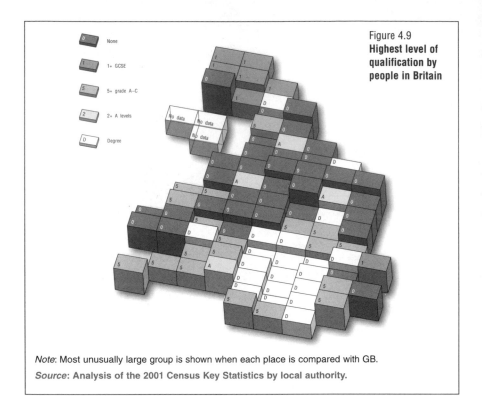

Figure 4.9
Highest level of qualification by people in Britain

Legend:
- 0 — None
- 1 — 1+ GCSE
- 5 — 5+ grade A–C
- 2 — 2+ A levels
- D — Degree

Note: Most unusually large group is shown when each place is compared with GB.

Source: Analysis of the 2001 Census Key Statistics by local authority.

What do people come to Britain for? Many come to work, but others come – and often pay dearly – for education, particularly university education. Just as within Britain access to a university determines many life chances, so too worldwide. This, however, is just a very small part of the explanation for the pattern of identity shown in Figure 4.9, the pattern that this book began trying to explain in Chapter 1, the pattern that the movement of children described in Chapter 2 results from, and the pattern that is partly the result of the education system described in Chapter 3 – the pattern of people's identities through their qualifications. The 2001 Census was the first Census to ask everyone about their qualifications, not just those with university degrees. What it reveals shows why education matters (for the sorting of children rather than teaching). If you had any doubt that education plays the major part in deciding where you end up living, just spend a minute looking at Figure 4.9. There are, of course, people with degrees living in every area of Britain (their numbers are in Figure 3.9), but within each area the pattern of location is as segregated as that shown nationally here. We are a nation divided by so-called knowledge.

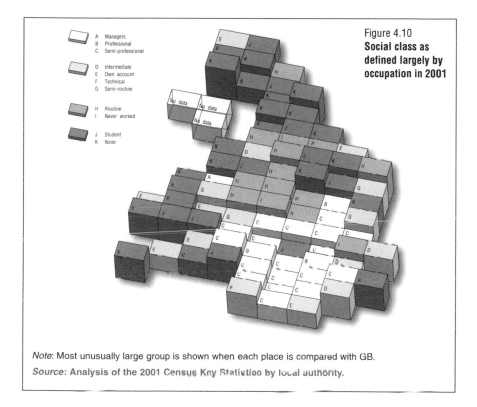

Figure 4.10
Social class as defined largely by occupation in 2001

A Managers
B Professional
C Semi-professional

D Intermediate
E Own account
F Technical
G Semi-routine

H Routine
I Never worked

J Student
K None

Note: Most unusually large group is shown when each place is compared with GB.

Source: Analysis of the 2001 Census Key Statistics by local authority.

The geographical link between the last figure and the final form of identity shown in this chapter (in Figure 4.10) does not require a great leap of imagination. Social class is the form of identity that in many ways social science began with. Classes are constantly defined and redefined and a new one was used in the 2001 Census Key Statistics, but no matter how they alter the definitions, the basic pattern prevails. People with no social class (under the new system which was envisaged to give everyone a class!) include those who have not worked recently: pensioners, many lone parents, people retired on grounds of ill health and so on. What is most interesting about this figure is not the extent to which it confirms our prejudices/understanding but what it tells us might be new and what might be the beginning of some major change. Consider, for example, the relative geographic isolation of the seven areas with unusual numbers working on their own account (only a minority will be farmers). And why is the only area with a preponderance of the highest class – managers – Cheshire East? Do the modern equivalent of mill owners need to crowd together for safety in the north? Why is one area of London typified by people who have never worked and where 165,000 people have degrees? See Figure 4.3 for part of the answer.

An exercise

(Preferably 84 players, but can be played with 336, 252, 168, 42, 21 or 10 and numbers in between with some careful organisation).

Before you begin assign each player to one of the 84 areas of Britain. If you have more than 84 players, work in groups. If you have less, each player is responsible for more than one area or, more easily, only use part of Britain. The instructions here are written assuming that you have exactly 84 players but it is not difficult to adapt them.

1 Each player needs to construct the hypothetical atypical individual living in their area using the information contained in Figures 4.1–4.10. If that place were a person, who would that person be? Note: this person is not typical of the area and may not even exist. Instead, they are typical about what is atypical of their area, what identities are more clustered there than in other places. For example, here is the atypical occupant of London Central:

   4.1   Female
   4.2   Aged 25–29
   4.3   Black African
   4.4   Muslim
   4.5   Single
   4.6   Living alone
   4.7   Many neighbours left the area since last year
   4.8   Born outside the European Union
   4.9   Has a university degree or equivalent
   4.10 Professional occupation

2 Once the players have identified their identities they need to move around the room or lecture theatre so that their 84 bodies are arranged as the map of Britain has been arranged in the Figures 4.1–4.10. They should have up to four neighbours, each neighbour corresponding to the area immediately North, East, South and West of them. Again you have formed a map of Britain.
3 Next compare your identities to those of your neighbours. On how many of the ten do you differ? For instance, to the East of the person who is London Central is London East and they differ on seven identities marked[*] below:

   4.1   Female
   4.2   Aged 0–4*
   4.3   Indian*
   4.4   Muslim (parents)

4.5   Married (parents)*

4.6   Married with kid(s) (that's you)*

4.7   People tend not to move home in the area*

4.8   Born outside the EU

4.9   Parents have no qualifications*

4.10   Intermediate occupation*

4 Now move slightly away from the neighbour(s) you share least in common with and slightly towards those who are most similar to you. Someone looking down on the room from above should see a geographical map of social and cultural divides and similarities opening up before their eyes.

5 To find where the greatest local divides in the country are someone needs to call out numbers: 'Is there any pair of people in the room who differ on all ten identities? On nine? How many on eight?', and so on. Where are these divides and is a pattern beginning to form to connect them? Are any two neighbouring areas identical in their identities? Are any pairs of non-neighbour areas identical in their identities?

6 Finally, start with the greatest local divides identified above and attempt to connect them using the next greatest adjacent divides to eventually divide the room and yourselves into two roughly equally sized groups. Your dividing line should ideally be continuous. It may become quite convoluted and complex as you attempt to link enclaves and exclude pockets that do not fit their original side of the line. It is advisable, and mildly amusing, to use a ball of wool to construct this line, starting with the pair of individuals who have least in common holding the wool in the middle and sending both ends out either way from between them to attempt to eventually join those two ends together again, having split the country in two.

With over 84 players it may be advisable not to attempt this last part of the exercise, especially if anyone needs to use the toilet in the near future.

If you are wondering what the purpose of the exercise is, it is to dispel a myth. It is often said by eminent social researchers that it is not possible to define areas as being rich or poor and to then expect them to contain most rich or poor people within them (or any other two groups). This is not true. It is possible. It is just that the dividing line you might have to construct, to be perfect, would be very long and very complex. It would snake up one street, taking in the odd house or two, then encompass an entire estate, save for a single person living in a flat in the middle, for whom a special loop would have to be drawn. The line would appear fractal in shape, like a coastline but even more convoluted than that.

Such lines are drawn everyday in the world. Rather like geo-demographics, there is a small industry that has developed to draw them. Political maps and voting districts are constantly redrawn, often in the interests of a particular group or party. This is known as gerrymandering, particularly where the attempt is to draw lines around voters to win seats. The job of gerrymanderers can be even more complex than yours when they try to draw lines around groups of people who are different from one another to ensure that the group they favour is just larger than the group they do not like. But more on them in the next chapter.

Finally, having geographically divided the country into two halves, compare the differences between areas either side of the dividing line to the differences between areas on the edge of one half and in its centre. Do areas either side of the line have more in common with each other than they do with other areas in their half? If so, what maintains the line of the line?

CONCLUSION

I have done what this chapter began by criticising. I have labelled places in ways that it can easily be argued are arbitrary. I have used a consistent method, but had I used different areas, different labels or different statistics about the same labels, the pictures would be a little different. Labelling is never precise. Places are mixtures as much as people are mixtures. Some men behave more like most women, some 30 year olds behave, or have the health of, say, a 40 year old. They say you are as young or as old as you feel. We are all a mixture of ethnicities, of origins. This is most true for people labelled White, whose ancestors are likely to have come from a wide range of areas, cultures, religions and to have spoken many languages – simply because rates of migration within Europe and from elsewhere were so high in the last few hundred years. None of us believes in a single, simply defined god. We all have our own beliefs and no two will quite match up. And yet people are willing to tick a box labelled religion. Similarly, people who did not tick the box did not for a wide variety of reasons. On a personal note, the religion I entered and defended in the newspapers was not recognised as a religious group despite outnumbering Sikhs, Jews and Buddhists on the Census forms. If the Census authorities don't like your answer, they delete it, as they did for gay people in 1991 and for many other groups before that. People are not simply single, married, divorced or widowed. The addition of remarried to the latest Census just begins to address this, but happily married, unhappily married, having an affair, two-timing my girlfriend, happy at home with the cats or living in a *ménage-a-trois* and endless other possibilities are all negated by labelling. Categorising people by households also does this. Being a pensioner living alone with no family can be very different from being a pensioner living alone in the house next door to your grown-up child. Similarly, we are all migrants and immigrants of different kinds, and we all have qualifications, most of which are not measured and for which you do not receive certificates. For instance, I cannot easily spell words over five letters long, but there was no box on the Census to let them know this. Finally, class is not a single simple ten-fold division of the population. There are a myriad facets to class. You don't either have the attributes of a class or not. All jobs are 'semi-routine', some just pay much better than others, involve a great deal more freedom and allow you to lord (a gendered expression note!) it over others more.

Why label then? Because people do. Because without labelling we could not begin to understand how the world works and, most importantly, because there is a little truth in the labels. Men do, on average, behave differently from women. This is a very large part of the reason why they are more likely to die young. Class matters – hugely, vitally, throughout your life and into the lives of your children. People are killed for believing in particular religions. People are treated very differently according to their skin colour (and again in Britain occasionally killed because of it). Labels matter. As long as we realise the damage they can do, the arbitrariness of their imposition and their intangibility and ephemerality, then we should use them. We have little choice. Everything, after all, is a label. For instance, each and every one of these words is one. You are labelling me as you read this 'bleeding heart, opinionated, liberal academic who doesn't know when to use only one word when he's given the space to use ten'! But how many labels to use and when to stop labelling…?

I could have written hundreds of pages on each map and for each map have drawn another hundred maps and written another hundred pages on each of them. However, as each additional map was drawn there would be less and less of substance to say. There is a simple, crude, true, describable, explainable, human geography to Britain. The majority of that quantitative picture is shown in the ten maps drawn in this chapter. After this, diminishing returns set in. The simple form of labelling used here is no cruder than any other, just more explicit. The patterns that you have just seen are the patterns that underlie almost all other maps of people's lives in this country – they are influenced by them and influence them in turn. Only 50 or so million people live here. Not quite a percentage of the world's population. It is a small island

and by far the most mapped island in the world. If we could not say with some certainty what people were doing, who they were, why they were here and where they are going, we could do so nowhere else on earth. Before you dismiss this form of stereotyping think about your life, the life of your fellow students and why you are all in the same room together. Ask your tutor where they came from. Will they not tell you? If so, why not? Perhaps it's because it will tell you a little too much about them?

5 Politics

...counting democracy, wasting votes

People are labelled by where they live using a variety of sources far wider than the Census and are, of course, themselves responsible for the labelling. In most cases, however, it is just a very small group of people who constructed the labels through which much of all our lives are described. In Britain these very small groups claim legitimacy in their labelling because either they are directly empowered to label others through the democratic process or the democratic process defends their rights to label. Thus democratically elected government defends and helps assert the rights of private companies, in a free market, to label populations by where they live. Similarly, researchers working in universities are both protected and encouraged by government to take part in such work, while many civil servants are directly employed by government to label places, most obviously through the construction of the decennial Census.

The labelling of places is part of democratically elected government in many ways. It enables government to target resources, for instance, or to see to what extent the nations it commands are moving apart or coming together. However, the same techniques that can be used to label places can also be used to examine how democratically elected governments come to be in power: to discern who votes for them, where their votes come from, how effective those votes are in returning their members to government, to see who is not voting or voting for other parties and how that is changing. The labelling of places is used to help put into power those who sanction and fund much of that labelling. It has long been argued that government needs to know, in detail, the characteristics of the populations it governs in order to govern them, in its terms, effectively. Over time surveillance of the population tends to increase. The fact that government also needs to know these characteristics to ensure that it is elected to govern is less widely appreciated.

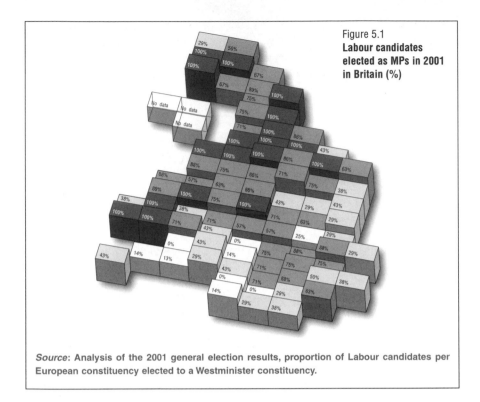

Figure 5.1
**Labour candidates
elected as MPs in 2001
in Britain (%)**

Source: Analysis of the 2001 general election results, proportion of Labour candidates per European constituency elected to a Westminister constituency.

In this chapter the geographical patterns of voting and political representation resulting from the British general election of 2001 are briefly described. This election was held just a few weeks after the Census of that year. The two had to be seen to be independent so the Census was not delayed despite the quarantine of many countryside areas due to the foot and mouth outbreak in cattle and sheep. There was much speculation that the government's handling of that outbreak would affect its pattern of votes and hence its majority in parliament, but in the event hardly any effect could be seen at all. The election returned the governing Labour party to power with a majority hardly altered from that it had achieved in its landslide victory four years earlier. Little appeared to have changed in those four years, yet when the geographical patterns are considered a great deal may well have altered.

Figure 5.1 shows the proportion of Labour MPs elected into power in each area of Britain. The European constituencies tend to be made up of either seven or eight Westminster constituencies (or 'seats'). Elections to the Westminster parliament are held on a first-past-the-post basis, where the candidate winning most votes in a seat wins that seat and the other votes are wasted. This is the most biased system of all the different electoral systems used in Britain, which a cynic may see as appropriate as it is that which elects representatives to the most powerful political body. In large parts of the north, Scotland and Wales, entire European constituencies are only represented by Labour MPs. Only in four areas of the south are there places represented by no Labour MPs. Of course, many people in the former areas did not vote Labour and a few people in the latter areas did vote for Labour. These people's democratically expressed wishes were ignored in the MPs sent to Westminster. The latter group can console themselves somewhat in that their votes may not have counted but their party did very well overall (as we shall see below). The first question to ask of this map is to what extent does the geography of political representation reflect the geography of political desires?

In each European constituency between 11% and 36% of the electorate (those eligible to vote) voted for Labour, a small minority in every case. In Northern Ireland no one voted Labour because they were not presented with that choice. The most pernicious bias in the British electoral system is the limitation of effective choice given by the very small number of candidates presented to each elector (very few voters are presented with an option to vote for a truly socialist candidate, for instance). Figure 5.2 does show that the pattern to the minority support for the Labour party does match that of its representation. This need not have been the case. With three or more parties it is quite possible for a party to receive the most votes in a European constituency but not to win any of the Westminster seats there. Incidentally, the share of the total vote Labour won varied between 16% and 66% in each of the constituencies above. The share of the total vote can be much higher because many electors chose not to vote where Labour did best. Nationally, Labour won 25% of the electorate, 42% of the vote, and 64% of the seats in Britain in 2001.

Proportionately, for every 1% of the electorate who voted for the Labour party that party won 2.6% of the seats in the 2001 general election. The first-past-the-post electoral system tends to award winners disproportionately in this way, but the disproportionality is far from geographically even. Figure 5.3 shows how the ratio of votes per seat varied across Britain in 2001. Each vote for Labour in the north and London was, in general, far more effective than were votes in the south. However, almost everywhere, Labour votes were more

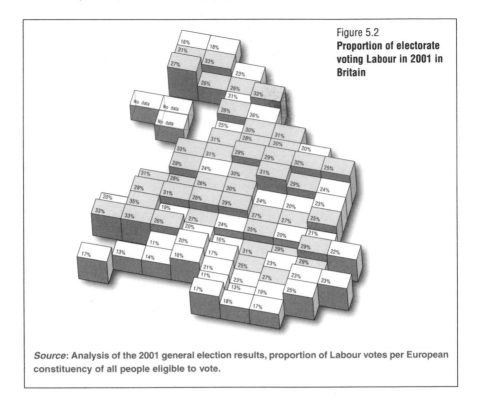

Figure 5.2
Proportion of electorate voting Labour in 2001 in Britain

Source: Analysis of the 2001 general election results, proportion of Labour votes per European constituency of all people eligible to vote.

effective than proportionality would suggest. Only in the four European constituencies where Labour won no seats and in three other neighbouring areas where they won a lower proportion of seats than the proportion of the electorate voting for them could these voters feel at least slightly aggrieved. In most of Britain Labour won all or almost all the seats in an area by winning just a few more votes than other parties but wining them in exactly those seats that mattered. In no seat did a majority of the electorate support the party which won such a large majority in parliament (the highest it won was 43% in Blaenau Gwent).

In case you were thinking that geography does not matter, consider Figure 5.4. In 1997 Labour won the support in Britain of 32% of the electorate and 65% of the seats; in 2001 they won only 25% of the electorate but 64% of the seats. Their vote to seats ratio rose from 2.06 to 2.59. Each vote for Labour was far more effective when cast just four years later but that pattern was not evenly spread across the country, as Figure 5.4 shows. In three areas

HUMAN GEOGRAPHY OF THE UK

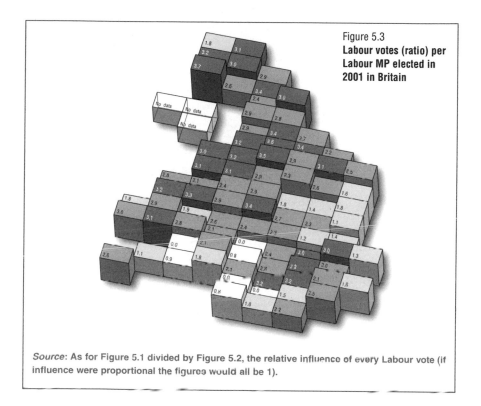

Figure 5.3
Labour votes (ratio) per Labour MP elected in 2001 in Britain

Source: As for Figure 5.1 divided by Figure 5.2, the relative influence of every Labour vote (if influence were proportional the figures would all be 1).

the efficiency of a Labour vote fell slightly but that was rare. Generally, the further north and west you travel, the more effective a vote for Labour became. This is how we came to the position whereby only a quarter of the population eligible to vote voted for the party which secured almost two-thirds of the seats. Every vote is equal, but some votes, most importantly those for Labour away from the South East (outside London), are very much more equal than others and are becoming more so. How did we get to the situation of government being supported by so few, and even those few having so little choice over which party to support? In most Labour seats only Labour can win. Where the seat is marginal, in most cases the only other party who can win are the Conservatives. This is not a great deal of choice, and is hardly embellished by the Liberal Democrats (and then in only a few places).

It would be very wrong to assume that support for the Labour party is now highest in the north and west of Britain and London. Figure 5.5 illustrates

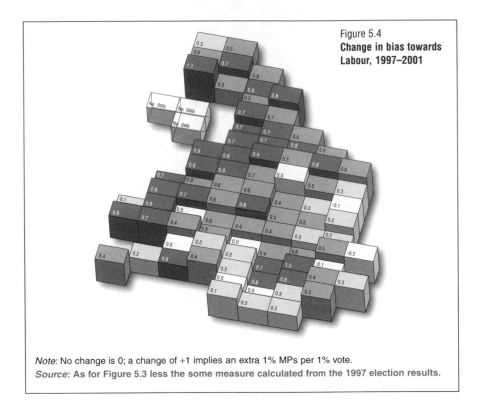

Figure 5.4
Change in bias towards Labour, 1997–2001

Note: No change is 0; a change of +1 implies an extra 1% MPs per 1% vote.
Source: As for Figure 5.3 less the some measure calculated from the 1997 election results.

that support is highest there for no party. In fact, a majority of electors in six areas shown in Figure 5.5 did not vote for any party. Nationally, 41% of electors did not vote in 2001 as compared to 28% just four years earlier. Above all else it is this rise in abstentions that accounts for the rise in bias that makes every remaining vote for Labour so much more valuable. Comparison of Figure 5.5 with Figure 5.2 shows that in every European constituency more people didn't vote than voted for Labour and the gap was highest in London Central where 27% more people abstained than voted for Labour. A majority of the electorate in 2001, as measured by first-past-the-post (or at least the single largest group as measured conventionally), is choosing to support no political party and this is the case everywhere and for all other political parties at the level of geography used throughout this book. Political bias is high because a growing number of people do not want to choose between what they are offered. They see so little point in the process.

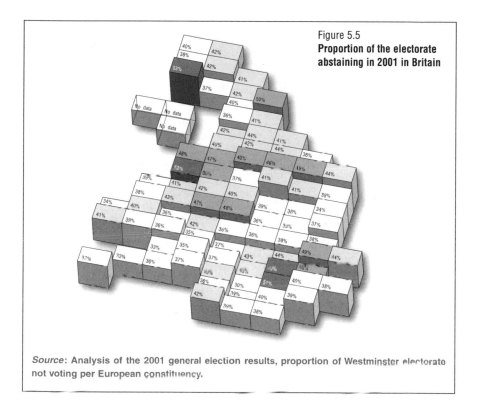

Figure 5.5
Proportion of the electorate abstaining in 2001 in Britain

Source: Analysis of the 2001 general election results, proportion of Westminster electorate not voting per European constituency.

Although the proportion of the electorate who chose not to vote increased everywhere, it was raised the most in the north west of England and in Scotland, where it was already very high. Figure 5.6 shows how abstentions rose most where bias increased most (Figure 5.4), albeit with some variation. Rising abstentions, given that they require the government to secure fewer votes for victory, are in the interest of the Parliamentary Labour Party in terms of securing its continued dominance of its government. Abstentions rose the least in those places where voters were most opposed to the Labour party and tried to use their votes to express this. Had they realised how the electoral system worked, they may have understood how futile such an attempt was and abstentions would have risen as much there too. In the run-up to a general election the electorate is bombarded with information suggesting that its vote counts immensely. Almost nothing is said about some people's votes counting much more than others and many being of no consequence at all. Despite this lack of

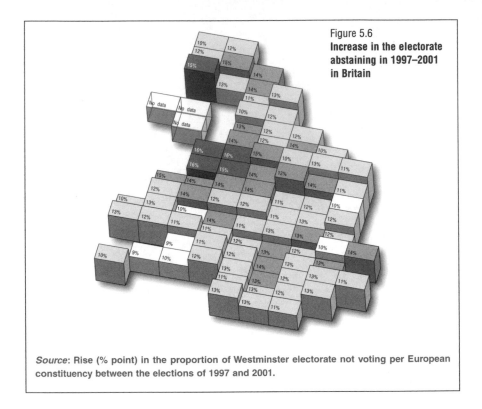

Figure 5.6
Increase in the electorate abstaining in 1997–2001 in Britain

Source: Rise (% point) in the proportion of Westminster electorate not voting per European constituency between the elections of 1997 and 2001.

information, more and more electors would appear to realise that for them, individually, voting may be futile.

Why vote at all if your vote carries so little weight when you don't vote for the ruling party? One answer is because it has not always been and will not always be like this. In 1997 the electorate of Britain voted out a Conservative government which had been in power for 18 years. There are many reasons as to why voters did this, but it is important to realise that popularity does not necessarily lead to power. For instance, in 1951 a majority of the electorate voted Labour but they won a minority of the seats. In 1997 the electorate both voted strongly and effectively for Labour. They voted most strongly, as they have always done, in those areas which are most disadvantaged by how life in Britain is organised. One example is the education system which is largely run by government. Figure 5.7 compares the proportion of children who achieved low qualifications at school before and just after that election with the proportion of adults (including many of their parents) who voted for Labour in 1997. The

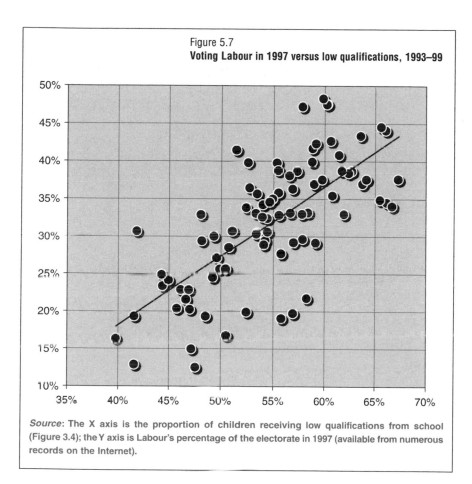

Figure 5.7
Voting Labour in 1997 versus low qualifications, 1993–99

Source: The X axis is the proportion of children receiving low qualifications from school (Figure 3.4); the Y axis is Labour's percentage of the electorate in 1997 (available from numerous records on the Internet).

Labour party, being the party which historically arose from the people who benefited least from the state, was the political party seen in 1997 as being most likely to rectify this injustice (among many others). Indeed, 'education, education, education' was one of the political slogans of that party. The Labour government, first elected to power in 1997, may well have helped improve the education system, or at least have stopped it becoming as unequal as it might have become, along with many other aspects of society. However, that government also failed to convince many of its supporters that it was doing so, which is part of the reason why its support fell so fast among the electorate. The electoral system and its particular geographical distribution of votes prevented that decline in support having any real effect.

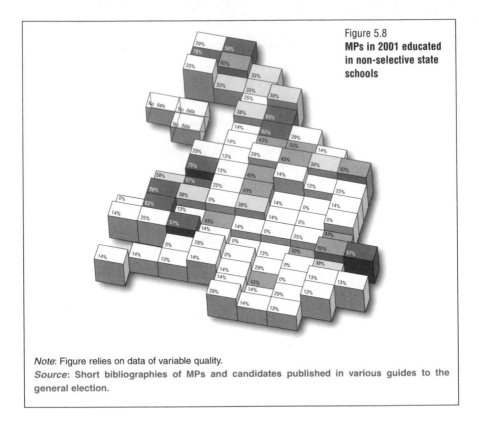

Figure 5.8
MPs in 2001 educated in non-selective state schools

Note: Figure relies on data of variable quality.
Source: Short bibliographies of MPs and candidates published in various guides to the general election.

Why should many of the electorate perceive government as not delivering, for example on education? One reason is that it is not easy to change a great deal in a short space of time. Another reason is that the impetus for change among MPs may not be too great. It is perhaps unsurprising that MPs might not see educational disadvantage as being really the most pressing of problems. Most of them have benefited from such disadvantage. Figure 5.8 shows the proportion of MPs elected to a seat in 2001 in each area who were educated in what appears to have been a comprehensive, secondary modern, technical or other generally non-selective state-funded school. In a majority of areas only one or no MPs out of the seven or eight representing the constituencies therein were educated as the majority of their constituents were. Only in ten areas were more than 50% of MPs so educated. Two-thirds of children are now taught in non-selective state schools (Figure 3.5). Less than one-third (177) of the MPs who represent them had similar backgrounds. Admittedly, 150 of those 177 are Labour MPs, but that is still only 36% of the

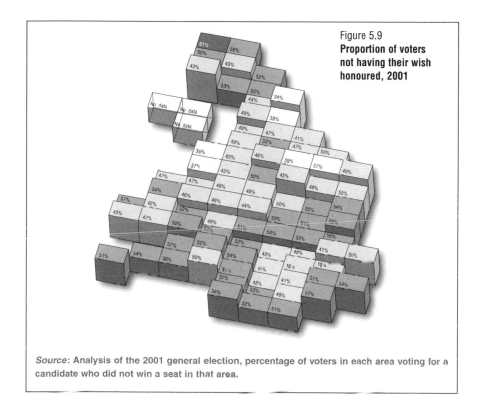

Figure 5.9
**Proportion of voters
not having their wish
honoured, 2001**

Source: Analysis of the 2001 general election, percentage of voters in each area voting for a candidate who did not win a seat in that area.

parliamentary Labour representation. It is also fair to say that most MPs were educated when the 11 plus was in place, but few younger Members of Parliament have not been to university. MPs tend to have more in common which each other than with their constituencies and as time passes and MPs' salaries rise this becomes ever more the case.

General elections are a remarkably good mechanism for annoying voters. Even after ignoring the largest single group of electors – those who chose not to vote – almost a majority remains of people who voted but who did not see the candidate they voted for win, some 12.5 million voters or 49% of everyone who voted (Figure 5.9 shows how little this varies across the country). Only 8.3 million or 32% of all voters voted for the Labour party and saw their candidate win in 2001. That leaves 68% of everyone who voted either seeing their preferred candidate lose, or win but their party fail to form a government (and so their elected representative has very little real power). It is difficult to think of an electoral mechanism better designed to annoy the majority of the electorate most of the

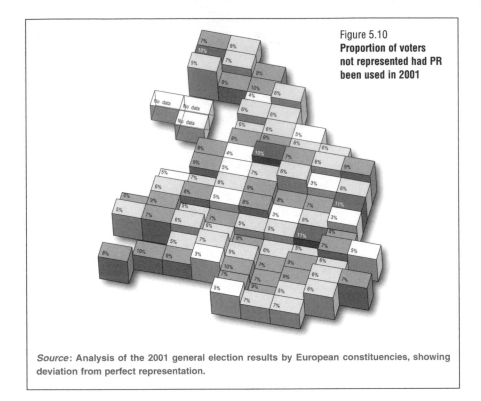

Figure 5.10
Proportion of voters not represented had PR been used in 2001

Source: Analysis of the 2001 general election results by European constituencies, showing deviation from perfect representation.

time, save that in the United States of America. Remember also that many of the 8.3 million people who voted Labour and saw their candidate win may well have done so as they saw the Labour candidate as the least worse alternative. In practice, less than a tenth of the entire population of Britain may actually be getting what they really want when it comes to democracy in this country.

The final map in this chapter shows the proportion of voters who would still not have had their wishes expressed through political representation had a simple system of Proportional Representation (PR) been used in 2001. As an example, take London Central's eight Westminster constituencies, which, under first-past-the-post, returned six Labour MPs and two Conservatives. Out of an electorate of over half a million adults, some 269,000 chose to vote in 2001. Of those, 47% voted Labour, 30% Conservative, 17% Liberal and 7% for other parties. If each of these four groups is allocated 4, 2, 1 and 1 seats respectively, then only 9.5% would have wasted their votes (30% − 25% + 17% − 12.5%) as compared to 49% (Figure 5.9). Figure 5.10 shows that there would have been no clear geographical pattern to these small proportions of wasted votes. Nationally, parliament would have been hung, the Conservatives would have won 206 seats, Labour 282, the Liberals 120, the Nationalists 22, and other parties 11. Between them, and irrespective of events in Northern Ireland, Labour and the Liberals could have formed a majority with 402 seats. The holder of the other parties' seat in London Central would have been the Green Party (which wins no seats under the first-past-the-post system).

An exercise

After the Labour government came to power in 1997 it redesigned the standard regions of Britain to create 11 Government Office Regions (GORs). Some of these have boundaries which have not changed for centuries (Scotland) or decades (London) but other regions are very new creations. These regions were created to be the areas for which regional governments would be elected. The first three such governments are in place in Scotland, Wales and London. Referendums may be held in three more regions in the near future to determine if they too should have such regional government: the North West, Yorkshire & the Humber and the North East. All six of these areas have the largest share of their electorate (who vote) voting Labour in 2001. Table 5.1 shows the number of voters in the general election voting for each party (Conservative, Labour, Liberal Democrat, Nationalist, Other), the electorate and the total vote. In one of the five regions which do not yet have regional government or a planned referendum, Labour received just over half a million fewer votes, in total, than the Conservatives.

The Labour party should win comfortable majorities in all the regions likely to have regional government in the near future. Because various systems of proportional representation are used in each of the regions which currently have regional government, they cannot usually expect to hold power but they can expect to be the largest party in each area. Given their current pattern of support, of the five regions remaining they can only expect to be the largest party in the East and West Midlands. Had the Labour government designed these regions a little differently they could have assured that they would have been the largest party in more of them. For example, Labour could have made the East of England Region a neater shape and one more fitting its label by placing its two most westerly areas in London (Hertfordshire and Essex West & Hertfordshire East) while placing the two most easterly areas of London in East of England Region (London East and London North East). They could have argued that Hertfordshire had more in common with London and that it was not, in general, a good thing to have a capital city as a single region. As a side

Table 5.1 Number of votes and the electorate (thousands), 2001 by Government Office Region

GOR	Con	Lab	Lib	Nat	Other	Electorate	Total	
1	908	1344	511	0	133	5216	2896	London
2	1404	947	776	0	136	5305	3264	South East
3	1052	715	843	0	108	4207	2719	South West
4	1087	969	454	0	103	4215	2612	East of England
5	848	1109	352	0	128	4176	2437	West Midlands
6	721	868	298	0	44	3171	1931	East Midlands
7	828	1456	476	0	95	5114	2855	North West
8	596	987	346	0	85	3566	2015	Yorkshire & The Humber
9	261	662	191	0	29	2009	1142	North East
10	289	667	189	196	32	2229	1373	Wales
11	361	1017	378	464	93	3981	2314	Scotland
Britain	8354	10740	4814	660	986	43187	25555	

Note: Numbers are thousands of voters or votes at the general election.

issue, had they done this, it would have moved enough voters between the two regions so that Labour had the largest share in both regions. To see how that would have occurred you need to use Table 5.2.

For the purposes of this game you can ignore areas 51 to 84 in Table 5.2. With the exception of North Yorkshire, they are all Labour areas. Only consider areas 1 to 50, regions 1 to 6. The remainder are included merely to provide a complete picture. To play the game you need to divide the class into six groups and assign each group to one of the six Government Office Regions. You then have to assume that you are all political party advisers attempting to devise a set of regions in the South of England in which Labour will win the majority of votes in the maximum number of regions. To keep some kind of balance you can only have six regions and no region may have less than five areas or more than ten areas. Each region must be made up of a set of contiguous areas (sharing a boundary in common). Start off with the current six regions. Here is how to play:

1 Each group needs to determine which area within its region, bordering another region, it would most like to lose to that region. If it has more Labour votes than it needs, it would make sense to lose an area with a high proportion of Labour voters in it. Similarly, each group should determine which neighbouring area in a bordering region it would most like to gain.

Table 5.2   Number of votes and the electorate (thousands), 2001 by European constituency

Area	Con	Lab	Lib	Nat	Other	Electorate	GOR	
1	80	125	46	0	18	539	1	London Central
2	95	145	26	0	10	508	1	London East
3	104	162	39	0	13	563	1	London North
4	60	152	37	0	19	528	1	London North East
5	93	155	35	0	7	504	1	London North West
6	126	96	77	0	8	508	1	London South & Surrey East
7	125	124	61	0	11	533	1	London South East
8	43	144	56	0	17	530	1	London South Inner
9	93	106	77	0	10	472	1	London South West
10	88	135	57	0	21	531	1	London West
11	153	83	70	0	14	511	2	Buckinghamshire & Oxfordshire East
12	142	96	90	0	21	564	2	East Sussex & Kent South
13	132	92	99	0	14	536	2	Hampshire North & Oxford
14	149	128	54	0	14	557	2	Kent East
15	146	136	48	0	10	554	2	Kent West
16	152	60	110	0	13	529	2	South Downs West
17	148	67	84	0	12	505	2	Surrey
18	139	89	67	0	15	508	2	Sussex West
19	120	105	67	0	10	506	2	Thames Valley
20	123	91	85	0	14	535	2	Wight & Hampshire South
21	103	138	85	0	14	535	3	Bristol
22	106	88	123	0	16	532	3	Cornwall & West Plymouth
23	147	71	127	0	21	543	3	Devon & East Plymouth
24	152	73	109	0	9	539	3	Dorset & East Devon
25	135	101	77	0	13	499	3	Gloucestershire
26	137	81	92	0	11	510	3	Itchen, Test & Avon

(Continued)

Table 5.2 (Continued)

Area	Con	Lab	Lib	Nat	Other	Electorate	GOR	
27	141	56	141	0	12	527	3	Somerset & North Devon
28	131	106	89	0	12	522	3	Wiltshire North & Bath
29	134	151	51	0	11	567	4	Bedfordshire & Milton Keynes
30	139	105	69	0	12	518	4	Cambridgeshire
31	135	109	66	0	14	523	4	Essex North & Suffolk South
32	123	112	36	0	11	501	4	Essex South
33	151	114	60	0	24	563	4	Essex West & Hertfordshire East
34	128	123	55	0	7	490	4	Hertfordshire
35	138	123	72	0	12	525	4	Norfolk
36	140	132	46	0	11	526	4	Suffolk & South West Norfolk
37	61	148	38	0	17	505	5	Birmingham East
38	93	139	36	0	13	528	5	Birmingham West
39	102	141	43	0	14	521	5	Coventry & North Warwickshire
40	132	106	70	0	44	548	5	Herefordshire & Shropshire
41	91	154	29	0	11	504	5	Midlands West
42	110	155	39	0	5	517	5	Staffordshire East & Derby
43	111	140	44	0	14	532	5	Staffordshire West & Congleton
44	146	126	52	0	10	520	5	Worcestershire & South Warwickshire
45	116	120	56	0	12	498	6	Leicester
46	143	132	54	0	5	553	6	Lincolnshire
47	146	148	46	0	9	549	6	Northamptonshire & Blaby
48	118	161	43	0	7	562	6	Nottingham & Leicestershire North West

(Continued)

Table 5.2 (Continued)

Area	Con	Lab	Lib	Nat	Other	Electorate	GOR	
49	81	153	55	0	3	494	6	Nottinghamshire North & Chesterfield
50	118	154	44	0	7	515	6	Peak District
51	104	134	45	0	6	489	7	Cheshire East
52	104	164	45	0	5	525	7	Cheshire West & Wirral
53	134	137	56	0	10	529	7	Cumbria & Lancashire North
54	62	127	69	0	10	529	7	Greater Manchester Central
55	53	142	51	0	20	495	7	Greater Manchester East
56	75	165	40	0	4	534	7	Greater Manchester West
57	107	123	45	0	13	497	7	Lancashire Central
58	105	156	36	0	6	504	7	Lancashire South
59	42	158	38	0	11	481	7	Merseyside East & Wigan
60	44	150	52	0	10	531	7	Merseyside West
61	99	134	52	0	12	530	8	East Yorkshire & North Lincolnshire
62	83	160	46	0	12	536	8	Leeds
63	137	101	80	0	12	514	8	North Yorkshire
64	44	131	58	0	8	449	8	Sheffield
65	51	165	32	0	13	510	8	Yorkshire South
66	83	160	42	0	14	537	8	Yorkshire South West
67	99	137	36	0	14	490	8	Yorkshire West
68	87	148	37	0	6	475	9	Cleveland & Richmond
69	60	191	51	0	7	527	9	Durham
70	74	156	66	0	7	504	9	Northumbria
71	39	167	36	0	9	503	9	Tyne & Wear
72	63	83	52	66	5	409	10	Mid & West Wales
73	77	135	36	46	5	480	10	North Wales

(Continued)

Table 5.2   (Continued)

Area	Con	Lab	Lib	Nat	Other	Electorate	GOR	
74	59	156	42	29	6	481	10	South Wales Central
75	52	161	31	28	8	467	10	South Wales East
76	38	132	28	27	7	393	10	South Wales West
77	27	183	27	70	16	552	11	Central Scotland
78	19	144	24	43	20	533	11	Glasgow
79	31	53	60	45	8	326	11	Highlands & Islands
80	52	140	59	51	12	542	11	Lothian
81	56	120	47	71	10	514	11	Mid Scotland & Fife
82	55	91	65	76	6	509	11	North East Scotland
83	72	132	54	55	9	509	11	South of Scotland
84	49	155	41	54	11	497	11	West of Scotland

Note: Numbers are thousands of voters or votes at the general election.

2   Each group then sends a delegation to each bordering region's group to negotiate whether its preferred choices are acceptable. It is permissible to accept sub-optimal transfers to areas if it can be argued that such a transfer is in the general long-term interest of the party.

3   To prevent the transfer of areas in one turn becoming too complex, each region may only lose up to two areas and gain up to two areas upon completion of all negotiations.

4   At no time can a region agree a set of transfers which would leave it with fewer than five areas or more than ten, or with any of its areas not being connected to all others within its region. Point-wise connections where two areas maintain contiguity at a corner are permissible under the United States version of this game, but not under UK rules.

5   Having successfully agreed a set of transfers, a halt is called to negotiations. All transfers occur simultaneously and each group then recalculates how many votes for each party its newly constituted region would gain if current voting patterns are assumed to prevail.

6   When recalculation has finished return to step 1 above, until the point at which no further areas are transferred between any groups.

7   Determine if any regions would now be lost by Labour and relabel the regions if their original geographical labels are no longer appropriate.

Note that negotiations can be complex. For instance, Hertfordshire in the Eastern region borders the regions of London, the West Midlands and the East Midlands. If the East Midlands group wishes to lose this area at any point (assuming it has been allocated

HUMAN GEOGRAPHY OF THE UK

there), it needs to determine whether moving any of these other regional boundaries would waste fewer Labour votes or could help to waste more Conservative votes and which move might be most acceptable to which other group. For example, 5000 more people vote Conservative than Labour in Hertfordshire, so moving Hertfordshire into London (by moving the London boundary north) wastes those extra Conservative votes.

Negotiations require some tact and it can be in the long-term interest of the party as a whole to accept a transfer that is not in the immediate interest of your group. All negotiation has to occur by persuasion. If you cannot persuade your colleagues in a neighbouring region that a particular transfer is in the general interest, then it cannot occur. It is a good idea to divide your group into negotiating teams to save time. For instance, if each group is made up of seven players, then three groups of two can be sent to, say, negotiate with small teams from three neighbouring regions, leaving one player to communicate between the three teams. It is sensible to set a time limit of about five minutes for each set of negotiations to be completed.

If the game is too easy, play it with all 11 regions. This is possible with only 11 players, but they have to work very hard. Under these rules, it is possible to create 11 regions in which Labour is the largest party in every area. If that game is still too easy, try splitting some of the largest regions in half and playing with more, smaller, regions (relaxing the minimum size rule). Eventually it becomes impossible to ensure that every region has a Labour majority, so what is the maximum number of regions that can be created to ensure a Labour majority in every region?

CONCLUSION

This is not a fanciful game. It is played out constantly in the United Kingdom, which makes more changes to its geographical boundaries every year than almost anywhere else in the world. However, instead of considering how 11 regions can be constructed from 84 areas, consider how thousands of local government wards can be created from even smaller areas when their boundaries are redrawn. Less frequently, constituencies themselves are redrawn using wards as the building bricks, over 600 areas being constructed from over 10,000 units. All this is largely done by hand and by negotiation. However, now imagine you are working to a more complex set of rules. Constituencies cannot cross county boundaries, must be almost equal in population size and you are no longer working towards a common goal but competing with other teams of negotiators working for other parties. Most difficult of all, under the official rules of the actual game in the UK, you cannot actually admit that you are playing this game. Instead you have to suggest that ward X should be moved into constituency Y because it has long-standing historical links with that area ever since farmer Z first took his cart to market in Y in 1066. Furthermore, you have to contend with a huge number of local mavericks who have their own particular interest in just one boundary or another.

It is rather like playing the game of Mornington Crescent. You are not allowed to admit you know what the rules are. And just like Mornington Crescent, most people do not realise what the rules really are – they are just very complex and secret. The worst thing you could do would be to try to claim there were no rules. That would be similar to suggesting that the United Kingdom had no written and understandable constitution...

6 Inequality
...income, poverty and wealth

Underlying most of the maps and social relationships discussed in this book are the inequitable geographical distributions of income and poverty. These patterns are long established. Some of the inequalities shown over the next few pages are similar to patterns seen several hundred years ago. What is new is the extent of inequalities in poverty, wealth and income in Britain now. The size of the gap between those who have a great deal of money and those who have very little is unprecedented, as are the sheer numbers of those living in poverty or on very high incomes. Such inequalities can become further entrenched as high incomes can be saved as wealth which in turn generates income. In Chapter 8 we consider some changes over time. Here we concentrate on the situation we find ourselves in at the start of this century.

Two sources of information are used in this chapter. The first is two reports from Barclays Bank, published in 2002 and 2003 on the Internet, profiling parliamentary constituencies in England and Wales according to the estimated incomes of Barclays' customers. These estimates were made from summing the steady streams of income entering customers' current accounts. Thus all forms of income are included and these figures relate to income often after income tax has been paid. Estimates have been made here for incomes in Scotland by establishing the relationship between the Carstairs deprivation index and the log of income and projecting figures for Scotland from that. For Northern Ireland no income or similar deprivation index was available and so data is missing from the province. Compared to Britain, incomes are generally lower and poverty is higher in Northern Ireland.

The second source of information used here are the components of the United Nations Development Programme (UNDP) Human Poverty Index for Britain by constituency, again published on the Internet but a little earlier than the Barclays data, in 2000. These components include the functional illiteracy rate of adults, people's premature death rate (on which Chapter 7 concentrates), long-term unemployment rate and the proportion of people living below half average incomes. Again these were published by parliamentary constituency and so can be aggregated to the European constituencies used in this book.

The two sources of information show much the same picture, but each from different ends of the income scale. Together they illustrate how poverty does not exist without affluence and affluence is not obvious without poverty. If there were very little poverty in Britain, the rich would be far less rich than they are. Without the rich there would similarly be far less poverty. For people to be rich others have to be poor, but not necessarily in the same places. Studying inequality requires studying both rich and poor simultaneously. While there is a great deal of information available on poverty, there is very little that has been released on affluence in Britain. Thus, to counter this, we begin with the highest income earners in the land.

In 2002 roughly 3% of Barclays' customers received an income of more than £60,000 a year, as estimated by the Bank. Comparisons with other sources of data suggest that this is a representative sample. Figure 6.1 shows how that proportion varies across the country, from 8% in London Central to 1.2% in Glasgow and the Midlands (West). In most of the country the proportion is less than 3%. In seven areas it is greater than 5%. An income of £60,000 may not sound very high to you, if you are from a wealthy family. Incomes tend to rise faster than inflation and high incomes rise over time more so again. Nevertheless this map is showing where the best-off people in Britain live in terms of their incomes, which are mainly from earnings but can include pensions and many other sources given the data source. Although the map is not showing the distribution of wealth, which we turn to briefly at the end of this chapter, a great deal of these incomes will be turned into various forms of wealth, most obviously through paying high mortgages and securing housing wealth, but also from contributing to what will be high pensions in the future, or paying for children's education to ensure they attend prestigious universities and become wealthy themselves. Look back over the maps in the preceding five chapters. To what extent were the patterns you were seeing there reflections of this geography of affluence?

In 2003 in each European constituency average income ranged from between £17,817 in Glasgow to £31,045 in London Central (Figure 6.2). The national average income of Barclays' customers was £21,851. Note that this average

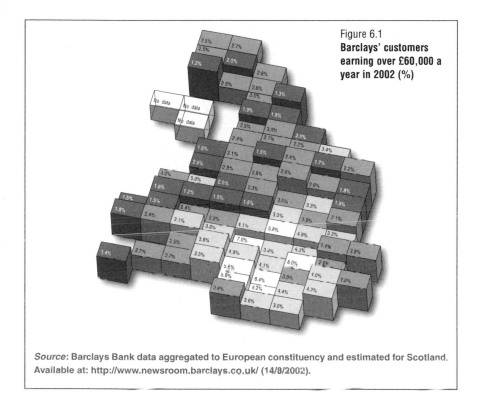

Figure 6.1
Barclays' customers earning over £60,000 a year in 2002 (%)

Source: Barclays Bank data aggregated to European constituency and estimated for Scotland. Available at: http://www.newsroom.barclays.co.uk/ (14/8/2002).

excludes people who do not have bank accounts but also those too rich to have normal bank accounts. The overall effect is that this income is quite close to officially estimated average household incomes in Britain. Again the geographical inequalities are stark. The country is cut in half roughly along the £22,000 mark. Average salaries in the far south west, south and east London and a handful of other areas in the south are low and just two areas of the north exceed this average. People are paid more in the south of England. Part of the reason for this is London weighting on some salaries, but that is only a small part of the explanation for these inequalities. Barclays Private Clients (BPC), who produced the survey, suggested another reason, which we will consider next. Incidentally, BPC is the 'wealth management arm of Barclays, serving one million clients in the UK and around the world in the areas of financial planning, investment and banking. BPC has some 25 per cent of the "big ticket" mortgage market (£250,000 plus) in England and Wales' (*source*: data press release). Their interest is in the rich.

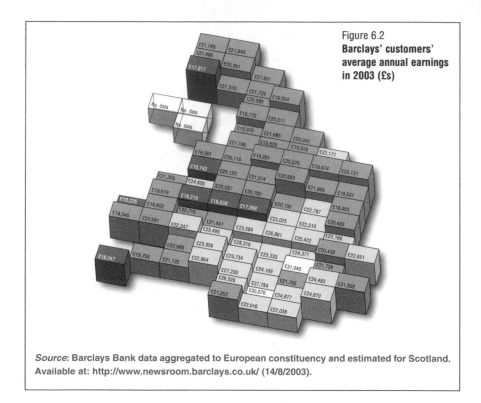

Figure 6.2
Barclays' customers' average annual earnings in 2003 (£s)

Source: Barclays Bank data aggregated to European constituency and estimated for Scotland. Available at: http://www.newsroom.barclays.co.uk/ (14/8/2003).

Figure 6.3 shows data Barclays released simultaneously to that shown in Figure 6.2 but now with average earnings roughly adjusted to reflect the regional cost of living. The comparable data for Scotland was estimated for this book using the regional weighting they included for Roxburgh and Berwickshire constituency (which they published as if it were part of England). What this map suggests is that once costs of living are taken into account London contains two of the poorest income areas in Britain. Most regions contain an area as affluent as London Central, and the Scots are particularly well off. The bulk of costs of living differences are housing costs. Although housing may be expensive for those who try to buy a house in the south when they are young, such costs are not wasted money. They are costs that are slowly being transferred to wealth. Thus this picture of income inequality is a little disingenuous. Nevertheless this is how affluent the average young person may feel in each area and thus also an indication of how rich a recent university graduate with a mortgage may feel shortly after securing a good job. Remember again though, that those in the south with mortgages are amassing considerable potential wealth while feeling temporarily poor.

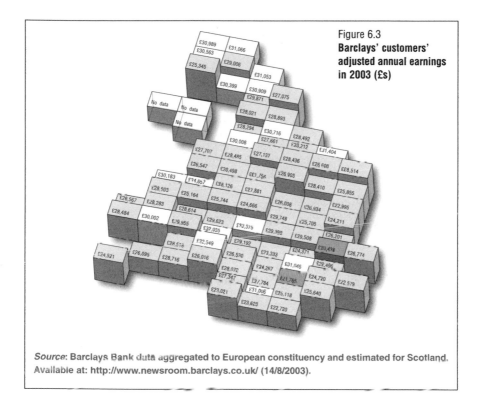

Figure 6.3
Barclays' customers' adjusted annual earnings in 2003 (£s)

Source: Barclays Bank data aggregated to European constituency and estimated for Scotland. Available at: http://www.newsroom.barclays.co.uk/ (14/8/2003).

Feeling poor and being poor are very different things. In Britain around 2000 roughly one in six people lived below the poverty line, as defined by the UN, having an income less than half the median average. Figure 6.4 shows where people on poverty incomes lived. In no area were less than a tenth of the population living on such low incomes; there are people living on poverty incomes everywhere. However, rates of poverty measured in this way are generally twice as high in the north as in the south and are clearly highest in Glasgow. There is a very strong relationship between this map and Figure 6.1, with one exception. In no area where more than 4% of the population earn over £60,000 does income poverty exceed 20%. In all areas where those living in income poverty exceeds 20% of the population, high earnings do not exceed 3%. The one exception is London Central, with both the highest proportion of high earners and a quarter of its population living below the income poverty line. In the centre of London the income rich and poor live very close together. Elsewhere, more often than not, the income rich pay not to live too near the income poor. There are many poor people living in London, and they are poor because their incomes are low.

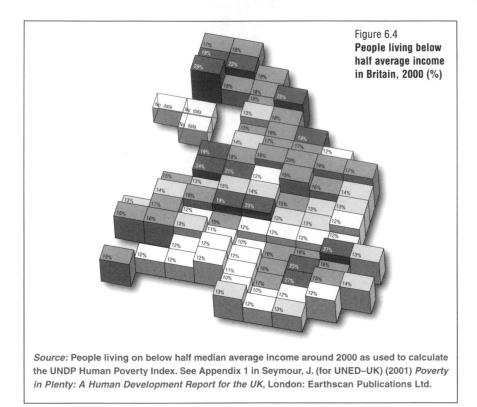

Figure 6.4
People living below half average income in Britain, 2000 (%)

Source: People living on below half median average income around 2000 as used to calculate the UNDP Human Poverty Index. See Appendix 1 in Seymour, J. (for UNED–UK) (2001) *Poverty in Plenty: A Human Development Report for the UK*, London: Earthscan Publications Ltd.

Figure 6.5 shows the relationship between mean average and the ratio of low to high earnings, combining data from the UNDP and Barclays. Average earnings can be predicted by the ratio of income rich to income poor people in an area in all areas except London Central. In Glasgow there are 25 people living on less than half national average income for every person living on £60,000 or more. At the other extreme, in Surrey, there are 1.2 people living on a poverty income for every one who is income rich. Put another way, to be able to estimate the proportions of people who are very rich and very poor you need only know average incomes in each area. The only exception to this generalisation is London Central. Incomes are particularly unevenly spread in London Central. Everywhere else the profile of inequality is similar. People are income poor because in every area incomes are unevenly distributed in similar ways. In poorer areas there are more poor people; but there are poor people too in affluent areas. The one exception to this pattern, central London, is a peculiar place where incomes are even more unequally distributed than

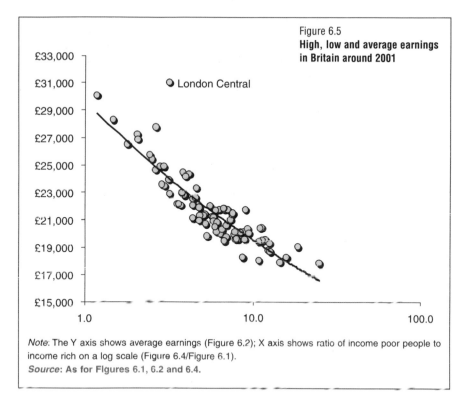

Figure 6.5
High, low and average earnings in Britain around 2001

○ London Central

Note: The Y axis shows average earnings (Figure 6.2); X axis shows ratio of income poor people to income rich on a log scale (Figure 6.4/Figure 6.1).
Source: As for Figures 6.1, 6.2 and 6.4.

elsewhere as there are more very rich people than average central London incomes would suggest, but also more very poor people living there. Elsewhere the story is simpler. There are high rates of income poverty in the north because there are low average incomes in the north. There are high rates of great affluence in suburban parts of London and the home counties because there are high average incomes there. Concentrations of poverty will, of course, reduce average incomes and concentrations of wealth raise them, but the relationships shown in Figure 6.5 do not simply imply that. Instead, they imply that income poverty is simply part of the spectrum of income inequality, as is income affluence.

When discussing Figure 5.6 on voting in elections, it was noted that: 'Although the proportion of the electorate who chose not to vote increased everywhere, it was raised the most in the north west of England and in Scotland, where it was already very high.' Consider now Figure 6.6, which

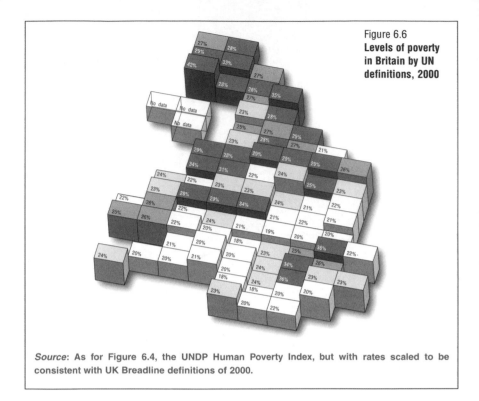

Figure 6.6
Levels of poverty in Britain by UN definitions, 2000

Source: As for Figure 6.4, the UNDP Human Poverty Index, but with rates scaled to be consistent with UK Breadline definitions of 2000.

depicts levels of poverty in Britain, and see just how similar the map of poverty is to the map of abstaining in elections. There are many ways in which poverty can be measured. To be poor means to have both low income and low wealth. As access to information on wealth is limited, the UNDP index uses a series of internationally available proxy measures to estimate levels of poverty. The rates have been scaled to agree nationally with the Breadline Britain measure of poverty – a consensual measure which involves asking a large sample of people what they consider to be necessities for life in Britain. The measure counts as poor those households which do not have access to those necessities and estimates their number. Around 2000 some 25% of households in Britain were poor by this measure, almost identical to the proportion of households (rather than people) living at below half average incomes (HBAI) after housing costs. Around 1991 the Breadline proportion was nearer 21%. For the 84 areas used here, the HBAI (Figure 6.4) and the UNDP index correlate with the Breadline index at 0.986 and 0.988 respectively. All maps of poverty look near identical no matter the measure used.

HUMAN GEOGRAPHY OF THE UK

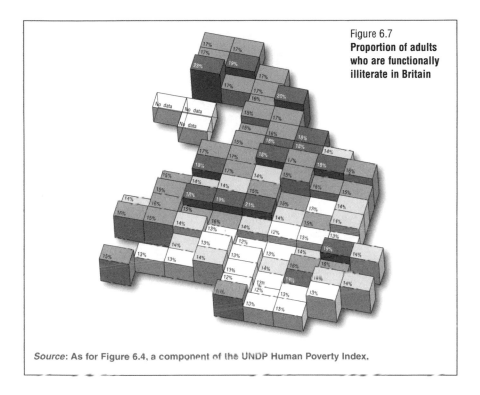

Figure 6.7
Proportion of adults who are functionally illiterate in Britain

Source: As for Figure 6.4, a component of the UNDP Human Poverty Index.

Not only do maps of poverty look identical to each other, but so many aspects of life are so closely connected to poverty that their maps appear near identical. Figure 6.7 shows the proportion of adults aged 16–65 who are functionally illiterate. Functionally illiterate means not being able to read and write at a basic level expected of people in the UK (see the source for details). All the areas of Britain where less than a seventh of people of these ages are functionally illiterate are in the south. The proportion is more than a sixth in all of Scotland and 15 other areas in the north, but only at this level in the south in two areas of London. Compare this map of adults to that of children in Figure 3.1. Most of the places where two or more children do badly at age 11 for every child which does well are also the places where adults in the past did poorly out of their education. Over time levels of literacy will rise, but there are no signs that inequalities will fall and average expectations will be higher in the future. Simply being able to read and write at a functional level will soon be seen as normal and not a problem for a minority of the population. Only a few decades ago it was not necessary for many people to be functionally literate to be able to lead a good life. In a few decades' time maps like this

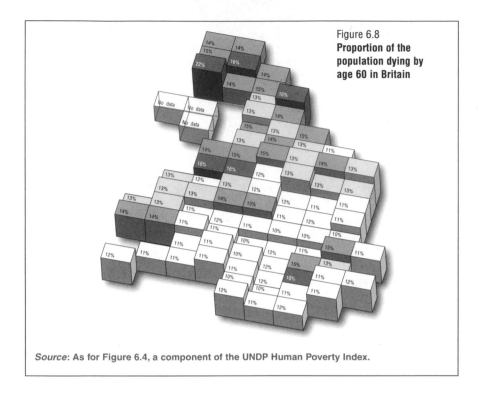

Figure 6.8
Proportion of the population dying by age 60 in Britain

Source: As for Figure 6.4, a component of the UNDP Human Poverty Index.

may be drawn of the information technology illiterate – and they may well look much the same.

Another component of the UN Poverty Index is premature mortality, measured here in Figure 6.8 as simply the proportion of people dying before their 60th birthday. By the end of the twentieth century over a fifth of people in Glasgow did not live this long and over an eighth of people in almost all of the north died before reaching 60. In the south such rates are only seen in four areas of London. All the components of the Human Poverty Index are related to each other. Premature death is not only an indicator of poverty but also one of its main outcomes. Not all forms of death are hastened by poverty, a few are even the products of affluence, as Chapter 7 illustrates, and there are many other factors which impact upon our health. However, growing up in poverty shortens lives as well as increasing children's chances of never learning to read or write properly, adults' chances of getting a good job and of earning or receiving a decent income. Lastly, Figure 6.8 shows where the people who do not feature on most of the maps and graphs described so far lived, that is

HUMAN GEOGRAPHY OF THE UK

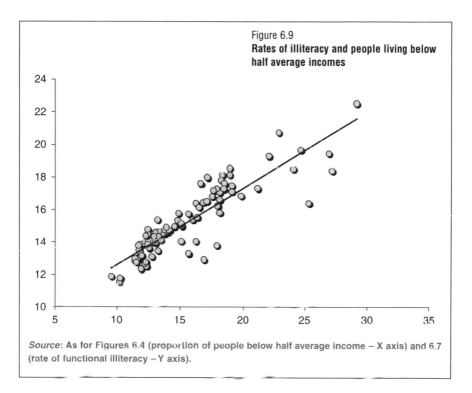

Figure 6.9
Rates of illiteracy and people living below half average incomes

Source: As for Figures 6.4 (proportion of people below half average income – X axis) and 6.7 (rate of functional illiteracy – Y axis).

the people who could have expected to be living in this country by the turn of the millennium, given when they were born, but who are not here because they have died young. The majority of those deaths are very strongly related to the human landscape in which they were living. Even in the affluent European constituencies the majority will have died within their poorer enclaves.

To illustrate just how close the relationship between these indicators can be consider Figure 6.9, which shows how it is possible to predict the proportion of adults unable to read and write properly from the proportion of house-holds living in each area on below half average incomes. Put roughly, in every area for every 100 adults, eight will be functionally illiterate, and a further one for every two households living below half average incomes. Of course this illiteracy is gained as a product of poor education in the past and so it could be claimed that it is the illiteracy, from that lack of education, which has led to low incomes. This will be partly true. But what led to the poor education of children in these areas in the past? Most areas that are poor now were poor in relative terms half a century ago. A uniform standard of education

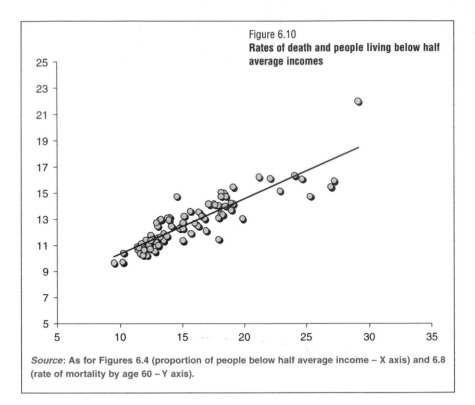

Figure 6.10
Rates of death and people living below half average incomes

Source: As for Figures 6.4 (proportion of people below half average income – X axis) and 6.8 (rate of mortality by age 60 – Y axis).

was not provided across the country then, as it is not now. Children from poorer backgrounds and areas then were much less likely to pass the 11 plus as compared to their more affluent peers. There were more grammar schools per pupil in affluent areas and the map of people's relative chances of getting a (albeit far fewer) university place was just as uneven, if not more so. There is a circularity reinforced by area whereby poverty in the past led to worse education, employment, health and housing which in turn all increase a person's chances of being poor in the future and then their children's chances of receiving relatively worse education, employment, health and housing, and on and on despite the general levels of these services improving. This process is amplified geographically most simply because the services are provided at an area level for groups of people, not for individuals, through schools, firms, the environment and the building and maintenance of settlements.

The direction of causality is perhaps a little less ambiguous when premature mortality is related to low incomes, as in Figure 6.10. In every area roughly

HUMAN GEOGRAPHY OF THE UK

seven people out of every hundred will die before age 60 and a further one for every three households living below half average income. However, again it could be argued that high rates of illness, preceding premature death, will result in fewer people working in poorer areas and thus in incomes being lower. Again though, you need to consider why more people were ill in these places in the first instance. Poverty begets poverty and, it can be argued, is, in aggregate over time, largely a product of the greed of the affluent. It is the affluent who tend to have the power to share resources more fairly among all children in a country; they not only tend to end up in positions of responsi bility but also actually have recourse to those resources most obviously through the wealth that they hold. It is the affluent majority who tolerate and maintain the organisation of society, the map of our human geography, which ensures that the children of poorer people are worse fed, educated, housed and employed. It should thus come as no surprise in the future that graphs showing even stronger associations than Figure 6.10 will be drawn. The precursors to premature death are being lined up more carefully geo-graphically than they were in the past. Why, if all this is so obvious, should the affluent behave like this? Well, for a start, it is in their short-term interest and their children's medium-term interests to behave like this. Secondly, as the following exercise illustrates, it is often not as obvious due to how this information is reported. Thirdly, people are only human: altruism, rational-ity, and decency are learned, not in-bred.

An exercise

Read the article below then follow the instructions after it:

Adding up to much less

Paul Foot, Wednesday 26 November 2003

The Guardian

On September 26, the leader of the House of Commons, Peter Hain, was on the BBC's *Newsnight* proclaiming the progressive reforms of New Labour. High on the list, he claimed, was the closing of the gap between rich and poor. "If you look at the figures," he said, "the bottom tenth of the population have seen their incomes increase by 15%, while the top tenth have seen their incomes reduced by 3%. That's redistribution."

This seemed so unlikely that I contacted the Office of National Statistics (ONS), now attached to the Cabinet Office. To my surprise, the figures it gave me confirmed what Hain said. The earnings of the poorest tenth, they revealed, had risen steadily since 1997, but, astonishingly, the earnings of the richest tenth, after growing even faster every year until 2001, suddenly and sharply went down in 2002.

I went back, twice, to the ONS, requoted the figures they had given me, expressed my doubts, and asked for an explanation. Back came the reply that the figures had been "double-checked" and were "correct".

No doubt about it, then. Peter was right and my scepticism was wrong. But wait. Early this month, the ONS sent me a new set of earnings figures, updated to 2003. They flatly contradicted the figures given to me previously. They showed a steady annual increase in earnings for both the poorest and the richest tenths at about the same rate from 1997 to 2003.

What was going on? I consulted Incomes Data Services, a specialist in such matters. Its explanation was rather shocking. "The ONS," it said, "has made a mistake. It has given you the upper quartile (quarter) figure for 2002 when it should have given you the highest decile (tenth)." The real figures for the richest decile showed a steady rise in earnings in every year from 1997 to 2003. So, under New Labour, the rich are getting richer and the gap between rich and poor is getting wider.

The ONS now admits its error, and has apologised to me. Where does that leave Peter Hain? His spokesman tells me that Hain's claim on *Newsnight* was based on a survey by the Institute of Fiscal Studies, whose press release concluded: "Focusing on (tax and benefit) measures that directly affect household incomes and spending shows a progressive pattern, varying from a boost of more than 15% to the incomes of the poorest tenth of the population to a loss of nearly 3% for the richest tenth". That says, vaguely and almost incomprehensibly, something rather different to what Hain claimed.

Much more specific and reliable are the latest figures from the Inland Revenue on the distribution of marketable wealth – which includes rent, dividends and other windfalls of capitalism. They show that the richest 1% of the population had 20% of the nation's wealth in 1996 and, thanks to Peter Hain and New Labour, 23% in 2001. This is a bigger, quicker leap in the booty of the mega-rich than anything achieved under any other postwar government, including Thatcher's. As for the poorest half of the population, they had 7% of the wealth in 1996. And after the first four caring years of New Labour, their share dropped – to 5%.

■

The newspaper article contains seven paragraphs, nine percentages, 13 dates and 538 words. Newspaper articles are often written 'from the top' so that the reader receives most information early on and need not read to the end. However, in this case, without reading to the end you cannot understand the beginning. Most readers do not read to the ends of most articles which they begin, especially ones which are complex, and this article is unusually complex although its author has written about as clear an account of events as is possible. Five people are involved:

1 Paul Foot – journalist who specialised in exposing corruption in the UK.
2 Peter Hain – MP and at the time leader of the House of Commons.
3 ONS spokesperson – nameless representative of the national statistics agency.
4 IDS specialist – nameless 'independent expert' from a private company.
5 Hain's spokesperson – nameless governing political party worker.

What do you believe motivated each of the actors in this story? To what extent are the various motivations a combination of overly suspicious minds, over-enthusiasm, incompetence, maliciousness, disingenuousness or other motives? For instance, does the rich becoming richer in terms of income necessarily mean that the gap between rich and poor is becoming wider? Does Paul Foot confuse income with wealth and with what significant implications, if any? With such questions in mind write a fictional letter to be published in the newspaper from the point of view of either Peter Hain, the ONS, the IDS or Hain's spokesperson either objecting to the piece or supporting it. Write the letter in pairs and then, as a class, act as the Letters Editor of the newspaper and select those three or four which read best for publication (vote on it). Remember that letters to newspapers tend to be very short and to the point. Now each write a fictional reply from the point of view of Paul Foot to one of the letters and select the letter that you think is most convincing as a group by vote. Finally, if you find this interesting, find out what has happened to trends in income and wealth in the UK since this article was written. If you can find this out, perhaps using the same methods which Paul Foot used, can you then write an article on it which is clear, well-written and could potentially be published?

CONCLUSION

Inequalities in income, wealth and poverty, their extent and whether they are rising or falling over time is among the stock-in-trade subjects of the *Guardian* newspaper which published Paul Foot's article. Few other newspapers take such issues so seriously and these are difficult issues to write about because many readers find arguments about changing unequal shares difficult to follow.

Given levels of functional illiteracy in the UK, many papers are written so that they can be read and understood by the average 12 year old or younger. Levels of functional innumeracy in the UK are far higher and so you will find only the most simple use of statistics in newspaper articles. Counts are always preferred over percentages. Anything more complex than a percentage is normally taboo. Often the detail of particular topical incidents, such as an MP's comments being allegedly misleading, is written about most verbosely while the key issue remains a footnote. For instance, regardless of what actually occurred during the first two terms of the millennial Labour government, the most important inequality is that given in the last sentence of Paul Foot's article – the sentence least likely of all to be read. Half the population of the UK have recourse to only a twentieth of the country's wealth, implying that the other half hold 95% of everything which can be given a monetary value. For half the population to have recourse to 19 times (95/5) more than the other half is the remarkably inequitable result of how we distribute our resources, the products of our labours, and chances in lives to our children. Yet this is an inequality we have become so accustomed to that we, journalists, statisticians and other commentators take it almost for granted. We are used to gross inequalities in wealth in society, but we are sufficiently embarrassed by them as to make it very difficult to obtain figures that can be mapped. A large part of the inequality is monies held in the form of housing wealth, but to map those alone would leave out the even more inequitably distributed forms of wealth. These include money itself (often called 'savings'), stocks, shares and other more illiquid forms of wealth which can and should be extended to future pension entitlements, insurance cover and future benefits from inequitable state support (for instance, the predictable financial value of attending university). Paul Foot's figures do not include many of these other forms of wealth, but even by his interpretation of the official statistics we are only 5 percentage points away from a situation in which half the population has nothing, or negligible wealth. The last time government held a Royal Commission to uncover the detailed distribution of wealth, in the 1970s, it concluded that half had practically nothing of value. But nothing was done to change that.

People and governments generally only become galvanised about such inequities during times of national crises, such as following the Second World War. Similarly, at the start of the twentieth century a Liberal landslide victory helped pave the way for many reforms which reduced inequalities in the first few decades then. The post-Second World War Labour landslide victory resulted in both a government and a change in underlying ideals which meant that for decades in the middle of the century levels of inequality were held historically low and in some cases reduced. The government in power at the time this book was written also won on a landslide, partly due to the rejection of a regime which had allowed inequalities to grow at their fastest rate in the last century since the Royal Commission reported. Whether this current government enacts policies which actually lead to any reduction in inequality has yet to be seen, as the *Guardian* article makes clear. It clearly partly wishes to, otherwise the leader of the House of Commons would not be making such boasts, but beneath the claims facts are harder to find. Any failure to reduce inequalities will result in the human landscape of the UK becoming every more ragged. The peaks and valleys of life chances will rise and deepen, cliffs become higher and slopes appear where there were once flat plains on these maps. Human landscapes change much more quickly than the physical landscape upon which their lives are played out, but they are similar in ways other than simply their appearance. Just as the physical landscape, at least at its surface, is largely made up of layer upon layer of sediment from past landscapes, so too underlying the landscape of the human geography of the UK are layer upon layer of the bodies from such past formations. As the human landscape changes its surface shape, layers of sediment are laid down beneath that. This sediment can be described in many ways. It is made up of the industrial and social past of society, of millions more lives than those currently on the surface. The simplest way into which to depict the sediment of society is perhaps to take another metaphor from physical geography – and look at our history from the point of view of our dead (and those who label them) and how their bodies came to be where they are, when they die. So now we turn from inequality to mortality.

Postscript
Sadly Paul Foot died relatively young in the summer of 2004. A record of his work was published in Private Eye *magazine in the autumn of that year and in numerous newspaper obituaries earlier.*

7 Health

...the sedimentation of society

When, and of what, will you die? It's eight in the morning although my body thinks it's 1pm. This book is now two years overdue to reach the publishers. I am sitting in a hotel room in Philadelphia in which the windows are sealed. The air conditioning is set to heat the room to 72 degrees Fahrenheit, which I have just realised is identical to the temperature of the tropical fish tank I have at home. Tropical fish in captivity tend to experience quite a high mortality rate, but set the temperature right, give them the right amount of food and light and they live longer. What tends to kill them is stress from swimming around in a confined overcrowded environment, diseases when new fish are introduced into their tanks, their own wastes when the filter system in the tanks don't work, and the actions of other fish. Their mortality rates reflect all these things. If there has been a power-cut at home and my partner notices it, then she can reset the circuit breaker and their water will continue to be filtered, heated and lit. If she does not, there will be a 'catastrophic event' and I will have to clean out the tank on my return. Tropical fish are not very different from people in the influence that their environment has on their health. They simply have a little less power to alter it, and their lives run out faster.

I have just lit the second cigarette of the day and am drinking the third cup of strong black coffee, wondering whether the hotel will charge me (or rather my university) for the coffee that appeared to be free in the room but is probably not. I am eyeing up the complimentary chocolate bar 'jam packed with even more peanuts', the main constituent of which is chocolate and wondering whether it would make a good substitute for breakfast. I do not have a particularly healthy lifestyle, especially for a middle-aged academic, and yet – if you believe the numbers – I am just as likely to reach three score years and ten (and perhaps a further five) as the average man of my age from my country is (if not much more if I continue to smoke). What I am squandering through my lifestyle are the advantages I gain through my job. The things I am worrying about are whether the coffee is free, why the window won't open and how the fish are back home. I just phoned the family to say I arrived okay, and they are fine. I thought it might sound a bit callous if I asked about the fish! My university have paid for me to attend a conference although no one will know if I am here or not. There are 58 concurrent sessions about to start; I will not be missed. I am being paid to write what I want, where I want, when I want. Compared to the vast majority of people walking below my sealed window, I have it easy. I can afford to go on holidays and get a great many compared to the Americans on the sidewalk below. I don't worry about the bills, heating or food. I have a great deal of control over my environment. Of course, I could be hit by a taxi walking out of the hotel this evening, I could suddenly feel a pain on the plane home and find out that my heart does not have long to work or I could have a very bad reaction to those peanuts. I could be diagnosed with cancer very early, in say a decade's time. But, odds on, I will die of a heart attack, in England, in relative comfort, in my early seventies. And there's a fifty-fifty chance I'll live a little longer. Life and death are not distributed fairly.

This chapter tells a story of how, in our mortality, we collectively lay down a record of our lives. Figure 7.1 shows the basic geographical distribution of mortality rates in Britain. The rates shown are direct age–sex standardised rates. What they show is the number of people who would have died in each area, given the mortality rates of that area between 1996 and 2000, if that area had had the same population demography as England and Wales had in the early 1980s (for which the rate is set to 1.0). This figure and the nine which follow are for all ages and both sexes. Thus, in Glasgow, people are 40% more likely to die in a given year than on average, 66% more likely than in the area with the lowest mortality rate: Dorset & East Devon. Imagine that each of the squares in the figure was not a place, but a fish tank. Then the further north and west you travel, barring two tanks in London, the worse the fish generally do. In general, it is the environments in these areas, these tanks, which result in this pattern, especially the past environment. However, in aggregate, people differ from fish in one crucial way. They can move between their tanks. The majority of people who die in Dorset & East Devon did not begin their lives there. They moved to that area later in life and were able to afford to do so. Conversely, the majority of people who began life in Glasgow left that city long before they died elsewhere in the UK or abroad. The lifetime migration of people amplifies the environmental inequalities between places.

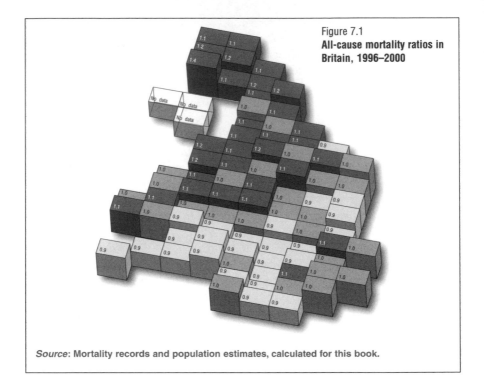

Source: Mortality records and population estimates, calculated for this book.

Within the overall geography of the dead, mortality by specific causes provides clues as to what it is about particular environments which leads to earlier or later deaths. In Figure 7.2, and the eight figures which follow, areas are highlighted where more than 25% additional deaths are due to a particular cause in an area or less than 20% fewer deaths than average are so caused. Variations are wider by cause than for all causes combined. The first cause of mortality shown in the figure here is an old disease: tuberculosis (International Classification of Disease categories, ICD9: 10–18, 119 and 137). It accounts for only 0.08% of all deaths in England and Wales by this period, two-thirds of the proportion 20 years ago. It is a disease that you are unlikely to contract if you are healthy, and is contracted by infection, which can largely be controlled to prevent mortality. However, despite all this a person is seven times more likely to die of tuberculosis in Glasgow as compared to the affluent enclave of North Yorkshire. In London and other urban areas in England rates remain above average as international migration brings in a steady stream of cases which were contracted abroad. In the poorest parts of Britain the disease continues to exist without the need for new introductions due to the poor

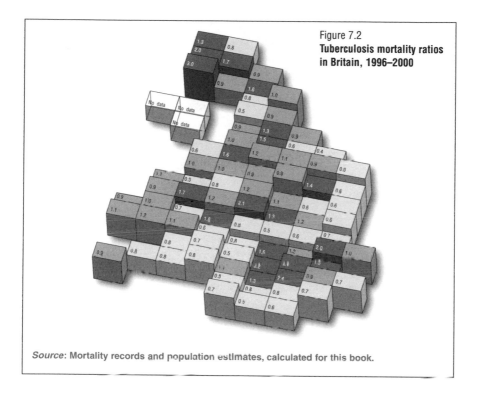

Figure 7.2
Tuberculosis mortality ratios in Britain, 1996–2000

Source: Mortality records and population estimates, calculated for this book.

environment for some people living there. This map is a tiny part of the jigsaw of death by cause which contributes to the overall rates, being what they are where they are.

Figure 7.3 shows the geographical distribution of a very new infectious disease: that associated with HIV (ICD9: 42–44, 279). Fewer people die of this than of tuberculosis, just 0.05% of all deaths, but almost seven times more than two decades ago when the disease was just beginning to be diagnosed. Infectious diseases in total account for only 0.61% of all deaths. They are of interest here because they illustrate just how important geographical location is to people's chances of contracting and dying of such diseases and because, when the next major infectious disease sweeps round the world, it is in these places that it is likely to kill most people first. For the rich western world, HIV turned out not to be the major pandemic once feared. Rates are highest in central London, Edinburgh, North East Scotland and East Sussex & Kent south (around Brighton). These are areas with high proportions of single people and a lot of migration, injecting drug users and high proportions of gay men. In each area the

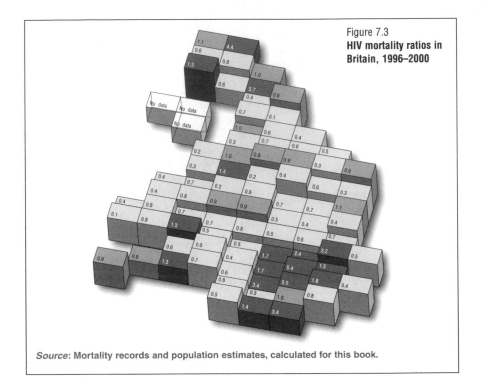

Figure 7.3
HIV mortality ratios in Britain, 1996–2000

Source: Mortality records and population estimates, calculated for this book.

specific immediate reason why rates of death from these diseases are unusually high or low is unique to that area. But it would be wrong to see these as the underlying reasons why rates are high in these areas. Before HIV diseases reached Britain none of these groups was at any risk of contracting them here (although of course they could easily contract them abroad). When the next new infectious disease arrives, the one thing we can be fairly sure about is that it will affect most where people mix most.

In contrast to infectious diseases, Figure 7.4 shows the distribution of deaths from lung cancer (ICD9: 162), which accounts for 5.4% of all deaths in England and Wales and 6.5% of all deaths in Scotland. This is a massive number compared to infectious diseases, accounting for almost ten times as many deaths as from all infections. It is a degenerative disease of which the major cause is smoking and yet the map opposite is a great amplification of the distribution of smoking. People's rates of smoking do not vary across the country as much as this map suggests and the rates varied even less in the past. In addition to rates of smoking, the map reflects the differing environments smokers lived in,

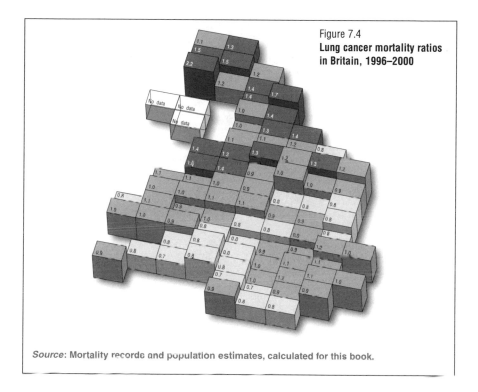

Figure 7.4
Lung cancer mortality ratios in Britain, 1996–2000

Source: Mortality records and population estimates, calculated for this book.

which made them more or less at risk to the damage of cigarettes. It also reflects the distribution of people working in industries where their environment could expose them to pollution that might cause this cancer. Most importantly it reflects the movements (or lack of movement) of people to where they are most likely to die. In the past, rates of smoking in London were high, but people left London when they were young and so the smokers of London have been spread across the south of England. People were more likely to move from north to south if they did not smoke, or to stop smoking if they moved south because there were fewer other smokers around them.

The sedimentary record of human life, as laid down in our deaths, is as much about our movements as it is about the places in which we lived. Figure 7.5 shows the distribution of deaths from skin cancer (ICD9: 172), which accounts for 0.26% of deaths in England and Wales and 0.20% in Scotland. The geography of skin cancer is clear and additional hours of sunshine in Devon, Cornwall and along the south coast will have contributed to this pattern, as will the overcast skies of the north and west. However, skin cancer can kill in

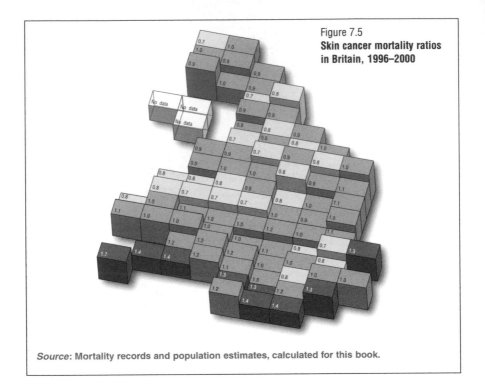

Figure 7.5
Skin cancer mortality ratios in Britain, 1996–2000

Source: Mortality records and population estimates, calculated for this book.

older ages and thus by the time people have made several migratory moves. Furthermore, it is likely to have been partly exposure to sunlight in much warmer climes than Britain which contributed to many individuals contracting this disease, particularly for former sailors but also keen sunbathers. Some of the pattern is of the people who come to live along the south coast of England who are both more likely to have been exposed most to sunlight and not to have died of another disease before that exposure could result in cancer. Younger people are more likely to die of the most fatal form of melanoma. Skin cancers, on aggregate, are diseases of the more affluent in Britain and so their geography partly reflects people's residential choices and their choices (made possible by wealth) in earlier years to be among the first generation to travel in large numbers abroad for their holidays. Maybe this is a large part of the reason why we see this particular pattern to these deaths.

The third cancer we consider here accounts for 0.21% of all deaths, less than one-hundredth of the 25% of all deaths which can now be attributable to cancers. Cervical cancer (ICD9: 180) only kills women and has reduced in impact to

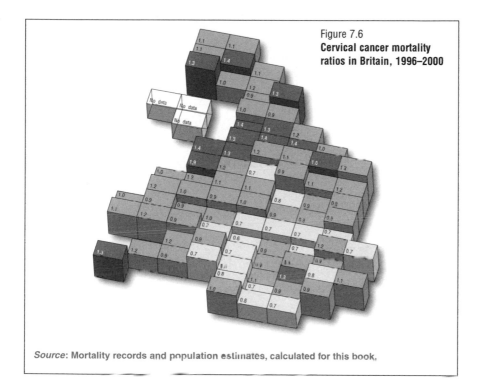

Figure 7.6
Cervical cancer mortality ratios in Britain, 1996–2000

Source: Mortality records and population estimates, calculated for this book.

almost 70% of its rate in the early 1980s, partly due to earlier screening and better treatment programmes. Figure 7.6 shows some very clear patterns to this disease. The low rates around the home counties ring are very distinct, as are the high rates in Yorkshire and the North West particularly. Unlike skin cancer, this is a cause of death which contributes to the overall geographical pattern of mortality rather than one which helps to reduce geographical variations. Again, patterns of migration are key to understanding the map seen here. People partly live in the home counties as a result of migrating into them over the course of either their or their parents' lives. People at lower risk of contracting cervical cancer or at a higher chance of being successfully diagnosed with that cancer early and being treated are more likely to live in such places. There are many factors which lead to some groups of women being more likely to have this disease than others, but just as those factors are important, so too are the factors that lead women with those risks to come to be living in particular places in Britain.

While cancers account for a quarter of all deaths, degenerative diseases of the circulatory system account for the largest proportion: 40%. Of these the

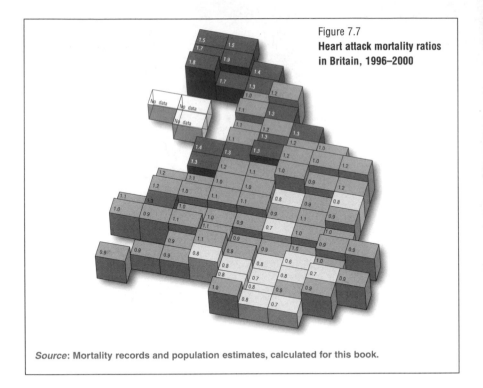

Figure 7.7
Heart attack mortality ratios in Britain, 1996–2000

Source: Mortality records and population estimates, calculated for this book.

single largest numbers are attributed to simple heart attacks (ICD 9: 410), responsible for 10.81% of all deaths in England and Wales and 14.56% of all deaths in Scotland by the start of this century. These proportions have been falling over time, but it will be many years before diseases of the heart are no longer the primary cause of mortality in Britain. Many factors make some people's circulatory systems more susceptible to diseases than others. Smoking and diet are important, yet Figure 7.7 shows patterns which yet again cannot simply be accounted for by such variations in behaviour. These are the major diseases which contribute to the overall geography of mortality in the country. Mortality rates in Scotland and parts of the north are simply too high to be purely a reflection of behaviour and social conditions there; rates in the south of London, through Hampshire to the coast are too low for local environments to simply be the cause of these patterns. Yet again we are seeing a distribution which is partly the result of lifetime migration. People who have been healthier during their childhood and working lives are more likely to have left the places with the highest rates. Such difference can be further amplified, as

HUMAN GEOGRAPHY OF THE UK

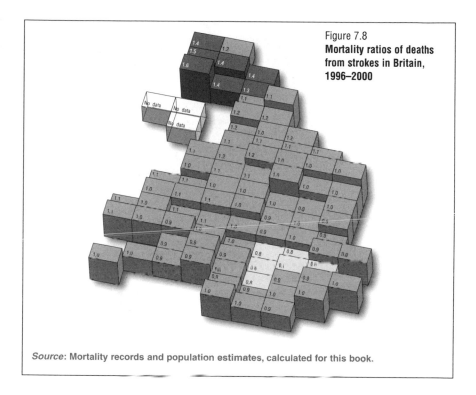

Figure 7.8
Mortality ratios of deaths from strokes in Britain, 1996–2000

Source: Mortality records and population estimates, calculated for this book.

in places where rates are lower, the health service is less stretched and better able to treat early symptoms of disease.

As people who are healthier in poorer areas have tended to move out of those areas, so too have healthier people in richer parts of the country tended to be among those most likely to move into the most affluent areas of those regions. Thus, just as for heart attacks, we see very clear patterns from deaths attributable to strokes (cerebrovascular diseases, ICD 9: 430–438), this collection of conditions being responsible for 10.28% of all deaths in England and Wales and 11.98% of all deaths in Scotland. Figure 7.8 shows how London's rates remain low, as London attracts ever more disproportionately the fit and those who are able to migrate there. However Scotland, suffering from depopulation of its more moveable people in many areas, is left with a population at far greater risk than average of dying from these causes. Of course, lifestyles in Scotland and London will differ, although more so now than they have in the past, and again rates of smoking are important. But if you look closely at

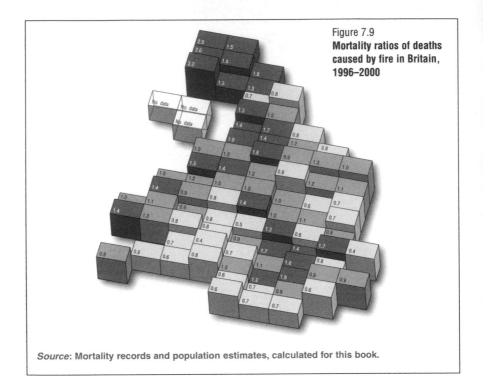

Figure 7.9
Mortality ratios of deaths caused by fire in Britain, 1996–2000

Source: Mortality records and population estimates, calculated for this book.

the figure, you will see that relentlessly as you move from north and west, to south and east, the rates of death from this disease fall. Environments matter, but they are made partly by the movements of people and cause people to move in particular directions. The deaths in these five years from stokes are laying down a very clear picture of many parts of the lives of the past people's of Britain, including how they came to live and die where they did.

Causes of death not related to disease are labelled as 'external' under the classification of mortality. This is a misnomer in that almost all the actual underlying causes of deaths are external to the body of the person who has died. Only 3.2% of all deaths are due to such causes, the majority of which are due to suicide, falls and motor vehicle accidents. Figure 7.9 shows the distribution of one cause outside this group – fire (ICD 9: 890–899) – which accounts for 0.07% of all deaths in England and Wales, 0.10% in Scotland. What is most interesting about deaths from fires in Britain in the context of this book is how they also reflect the overall pattern of mortality which is most clearly related to the distribution of poverty, and the patterns of migration which results from that

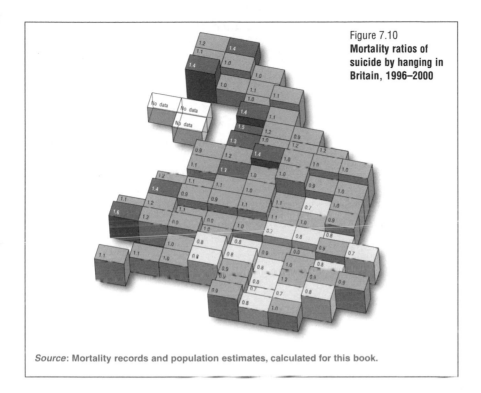

Figure 7.10
Mortality ratios of suicide by hanging in Britain, 1996–2000

Source: Mortality records and population estimates, calculated for this book.

distribution. But there are always caveats to such generalisations. Most deaths from fire are due to smoke inhalation of people who are unable to escape a fire or who are unaware of it. Figure 7.9 largely reflects the distribution of poverty, where people are more likely to live in homes without smoke alarms, with dangerous wiring and, yet again, where more people smoke. The most immediately dangerous fumes of a cigarette are from the fire it can light (thus they can kill in seconds as well as years). However, in addition to smoking, the map is also strongly influenced by where people live in flats. Thus even in affluent parts of London, rates are high where the population is crowded.

Figure 7.10 shows the last of nine selected causes of death chosen here to illustrate how each either adds or detracts from the overall pattern to death in Britain and how many factors influence each individual distribution. Almost 1% of all deaths are due to suicide or are 'undetermined accidents' (which are most likely to be suicide). There are many ways in which people can kill themselves, but the most common method, particularly among men, is by hanging, which accounts for one-third of all suicides. When the map is considered, again we see the home counties ring: people who get here are less likely to kill themselves, especially in this way. It is in the periphery of Britain where rates are high, at the edges of Wales, in the north west and in urban and remote Scotland. Suicide is perhaps the simplest example to give of a cause of death influenced by people's environment. There are very many reasons why people may seek to harm themselves, and why a proportion of those may manage to kill themselves, but in aggregate they reveal the geography that shows where some are more than twice as likely as others to resort to a ligature to hasten their own deaths. Thus, even how people choose to kill themselves has a geography. In the major cities, poisons and drugs are more common methods. In affluent areas, those fewer numbers who do resort to suicide are more likely to use the exhaust fumes from their cars.

An exercise

Table 7.1 lists the 100 major causes of mortality in England and Wales in the period 1996–2000, their ICD codes, disease label and the cumulative chance of dying of each cause of death. Thus a person has a 0.06% chance of dying of an intestinal infectious disease 'other' (other means not one of those listed below it, the ICD code for that cause of death is 8). The chance of dying of either that cause or tuberculosis is 0.14% and so on up to 100% when all diseases are included. The first column of the table is the last cumulative probability multiplied by 32,768 and expressed as a binary number. We can use the first column, and a coin, to give each member of the class a cause of death at random, assuming that the patterns in the future are similar to those in the past. The distribution of causes of death in the class should then reflect those in society as a whole.

To play the game each student needs a coin. Heads are '1' and tails are '0'. They will need to toss their coin up to 15 times to determine their allotted cause of death. In practice just a few tosses of the coin might be required however, and are required in most cases. For instance, suppose the first four tosses result in 'tails' followed by three 'heads'. Your number begins '0111'. Reading down the list of numbers (which is in ascending order) this is greater than 011011010101010, but less than 100010100010011 and so the cause of death is 'circulatory – heart attack'. If, however, the student tossed seven heads in a row ('1111111'), they would still need further numbers to determine their cause of death. If the next coin landed on heads, then the cause would be 'none recorded' (a few people are recorded as having died with no known cause). If it landed on tails, the student would need to toss the coin again and possibly again another two or three times to determine which kind of external cause they were to die from. Further tosses of '0100' would imply the student had been shot (by someone other than themselves).

HUMAN GEOGRAPHY OF THE UK

Table 7.1 One hundred major causes of death in England and Wales, 1996–2000

Chance	Number	ICD9 code	Disease	Chance (cumulative) %
000000000010011	1	8	infections – intestinal other	0.06
000000000101101	2	10–18, 119, 137	infections – tuberculosis	0.14
000000001111000	3	38	infections – septicaemia	0.37
000000010001001	4	42–44, 279	infections – HIV disease	0.42
000000010010011	5	70	infections – hepatitis	0.45
000000011001000	6	1–35, 37–139 remaining	infections – other	0.61
000000100101011	7	140–149	cancer – mouth	0.91
000001010001100	8	150	cancer – gullet	1.99
000010000000101	9	151	cancer – stomach	3.14
000011001011101	10	153	cancer – colon	4.97
000011101111001	11	154	cancer – rectum	5.84
000011111101100	12	155	cancer – liver	6.19
000100101001010	13	157	cancer – pancreas	7.26
000100101110111	14	161	cancer – larynx	7.40
001000001100110	15	162	cancer – lung	12.81
001000010111100	16	172	cancer – skin	13.08
001001101111011	17	174, 175	cancer – breast	15.22
001001111000001	18	180	cancer – cervix	15.44
001010000001110	19	179, 182	cancer – uterus	15.67
001010011111011	20	183	cancer – ovary	16.39
001011011110110	21	185	cancer – prostrate	17.94
001100000000000	22	188	cancer – bladder	18.75
001100010100101	23	191	cancer – brain	19.26
001110010100110	24	195–199, 159, 235–239	cancer – unspecified	22.38
001110101111000	25	204–208	cancer – leukaemia	23.03
001111011110001	26	200–203	cancer – lymphatic	24.17
010000010011001	27	140–239 remaining	cancer – other	25.47

(Continued)

Table 7.1 (Continued)

Chance	Number	ICD9 code	Disease	Chance (cumulative) %
010000011110001	28	240–249, 251–278	blood – endocrine not diabetes	25.74
010001001001111	29	250	blood – diabetes mellitus	26.81
010001011000001	30	280–289	blood – diseases of	27.15
010010001100100	31	290	mental – dementia, organic psychotic conditions	28.43
010010010011101	32	291, 303, 305, 860	mental – due to alcohol	28.61
010010011100011	33	304, 850–858	mental – due to drugs	28.82
010010101011101	34	292–302, 306–319	mental – other	29.19
010010101110111	35	36, 320–322	nervous – meningitis	29.27
010011000010111	36	332	nervous – Parkinson's	29.76
010011001011111	37	335	nervous – motor neurone and other	29.98
010011010001010	38	340	nervous – multiple sclerosis	30.11
010011010111100	39	345	nervous – epilepsy	30.26
010011110110101	40	320–389 remaining	nervous – other	31.02
010100000011010	41	390–398	circulatory – rheumatic heart disease	31.33
010100011010101	42	401–405	circulatory – hypertensive disease	31.90
011011010101010	43	410	circulatory – heart attack	42.71
100010100010011	44	411–414, 429	circulatory – chronic heart disease	53.97

(Continued)

Table 7.1 (Continued)

Chance	Number	ICD9 code	Disease	Chance (cumulative) %
100011001101001	45	415–417	circulatory – pulmonary circulation	55.01
100100111000101	46	420–423, 425–428	circulatory – other heart disease	57.63
101011011101100	47	430–438	circulatory – cerebrovascular disease	67.91
101011101001000	48	440	circulatory – atherosclerosis	68.19
101100101111010	49	441	circulatory – aortic aneurysm	69.91
101101110010000	50	390–459 remaining	circulatory – other	71.53
101101110101010	51	466	respiratory – bronchitis	71.61
101101110111111	52	487	respiratory influenza	71.68
110100011001011	53	480–486	respiratory – pneumonia	81.87
110111010111100	54	490–492, 494, 496	respiratory – chronic lower diseases	86.51
110111100001100	55	493	respiratory – asthma	86.76
110111101000011	56	500–508	respiratory – industrial lung disease	86.92
111000011111100	57	460–519 remaining	respiratory – other	88.27
111000111101010	58	531–534	digestive – ulcer	89.00
111001011101101	59	571	digestive – chronic liver disease	89.79
111010111011011	60	520–579 remaining	digestive – other	92.08

(Continued)

Table 7.1   (Continued)

Chance	Number	ICD9 code	Disease	Chance (cumulative) %
111011010111111	61	580–594	disease of kidney and urethra	92.77
111011101111011	62	595–629	genitourinary	93.34
111011101111101	63	630–676	pregnancy and childbirth	93.35
111100010010000	64	680–739	tissue, skin, musculoskeletal	94.19
111100010010111	65	740–742	congenital malformation – nervous	94.21
111100010111100	66	745–747	congenital malformation – heart	94.32
111100011010111	67	743–744, 748–759	congenital malformation – other	94.41
111100011011110	68	760–787	conditions in the perinatal period	94.43
111100011100010	69	780–796	other – signs and symptoms	94.44
111101110100011	70	797	senility without mention of dementia	96.59
111101110110101	71	798	sudden death, cause unknown	96.65
111101111100110	72	799	ill-defined and unknown causes	96.80
111101111101001	73	800–807	external – railway accidents	96.81
111101111110011	74	813, 826	external – pedal cycle accidents	96.84
111110000100001	75	814	external – pedestrian and motor vehicle	96.98
111110010011100	76	810–812, 815–825	external – motor vehicle accidents other	97.35

(Continued)

Table 7.1 (Continued)

Chance	Number	ICD9 code	Disease	Chance (cumulative) %
111110010011110	77	830–838	external – water transport accidents	97.36
111110010011111	78	840–845	external – air accident	97.36
111110010100100	79	861–869	external – accidental poisoning other	97.38
111110010111000	80	870–879	external – during surgery, medical care	97.44
111110110100001	81	880 889	external – falls	98.15
111110110110111	82	890–899	external – fire	98.22
111110111000001	83	901	external – hypothermia	98.25
111110111000011	84	904	external – hunger, thirst, exposure, neglect	98.25
111110111001111	85	910	external – accidental drowning	98.29
111110111011110	86	911	external – choking on food	98.34
111110111100000	87	919	external – caused by machinery	98.34
111110111100011	88	925	external – accident-electric current	98.35
111111000110101	89	950, 980	external – suicide/ accident by poison	98.60
111111001011001	90	951–952, 981–982	external – suicide/ accident by gases	98.71
111111011000110	91	953, 983	external – suicide/ accident by hanging	99.04
111111011010110	92	954, 984	external – suicide/ accident by drowning	99.09

(Continued)

Table 7.1 (Continued)

Chance	Number	ICD9 code	Disease	Chance (cumulative) %
111111011011110	93	955, 985	external – suicide/ accident by firearms	99.12
111111011100011	94	956, 986	external – suicide/ accident by cutting	99.13
111111011101011	95	957, 987	external – suicide/ accident by jumping	99.16
111111100100101	96	958–959, 988–989	external – suicide/ accident – other	99.33
111111100100110	97	965	external – assault firearms	99.34
111111100101011	98	966	external – assault cutting	99.35
111111100110100	99	960–964, 967–969	external – assault other	99.38
111111101101101	100	800–999 remaining	external – other	99.55
111111111111111	0	0	no cause recorded	100.00

This game may appear a little complex but all that it involves is in effect turning up to 15 tosses of a coin into a number between 0 and 32,767 to give a probability which can then be used to allocate a cause of death from Table 7.1. By reading down the binary numbers a digit at a time it quickly becomes apparent when no further tosses of the coin will alter the cause ascribed. The more rare a group of causes, the greater the number of coins which need to be tossed to first end up in that group and to secondly determine which cause it will be within that group.

Having allocated each member of the class a cause of death at random, next work out which groups of causes are most numerous in your population – infectious diseases, cancers, diseases of the blood and so on. For the most common causes, see if you have people allocated similar causes within those groups. It is important to remember that these causes have been allocated at random. They mean nothing for the people specifically allocated a cause. Some congenital causes only kill young babies, for instance, and several causes largely only apply to either men or women. However, for the group as a whole the distribution should be interesting. Here are a series of questions you can ask:

1   Are there causes that people are concerned about, for instance murder or air accidents which no one in your class has been allocated? Is this because they are rare or because of chance? If you think it is chance, try another random allocation of the class. If you are very good at maths, work out how many allocations you would have to make, given the size of your class, on average, until those causes were allocated?
2   What will kill the bulk of students in your class if they are representative of the population of England and Wales and if future causes of death are distributed as they are now?
3   Can you think of any reasons why your actual causes of death may be different from those allocated by this procedure, even if future causes of death are distributed as they are now?
4   Which of the causes that have been allocated do you think students will be less likely to die of in the future and which more likely, and why?
5   Given the maps above and the location in which you are playing this game (if that is in Britain) how might your local geography influence these chances?
6   Finally, if you could choose how you were to die in the future, what cause would you choose and why?

CONCLUSION

In aggregate, people leave more messages from their deaths than each individual mortality does alone. The deaths of the people of Britain lay down the sediment of its human geography, sediment which reflects in aggregate almost everything about their pasts. In the future, if disasters and wars do not have a great impact, the population will live longer than it has ever done. However, the spatial patterns of people's deaths may be even more distinct than those shown in the whistle-stop tour of mortality presented in the last few pages. The determinants of premature mortality and longevity are becoming more spatially distinct. From people's conditions in childhood and around the time of their birth, through to their behaviours as adults, the kinds of job they do and the rewards, freedoms, pressures and threats which they face, through to their recreational and retirement opportunities, the maps of this country are changing and those changes will be reflected in the patterns of future deaths. Most importantly, migration patterning is becoming ever more distinct. People are sorting themselves out in space, by place, more and more keenly as every year passes, as house prices diverge, and as yet greater proportions of the population leave home to attend university. Above all else, two forces, poverty and migration, create and amplify the spatial patterns revealed in our mortality. They also strongly influence our collective behaviours which are the immediate precursors of some of the biological causes of our physical deterioration. Although we have no accurate figures on the geography of smoking, those estimates that have been made suggest that the map strongly mirrors that of poverty. Of course there are exceptions, but they are becoming less and less evident, and as social groups are becoming ever more corralled together through their differential migration, they increasingly conform to what is normal for that place and that group.

A few minutes ago I took a break from writing this chapter. I went downstairs and had my late breakfast and then I stepped out of the hotel for a smoke. There are over 5000 delegates at this conference. Only one other person was smoking on the pavement with me from inside the building. Outside many of the people going about their normal business had cigarettes in their hands: the bus-boys on their breaks, the secretaries on an errand, the taxi drivers standing by their cars waiting for the next fare. Twenty-first-century America can partly be seen as a model for what twenty-first-century Britain will soon be. The middle-aged, middle-class academics are going to live for a long time. Their hearts will be stronger, they will be less susceptible to certain cancers and they will be among the first to benefit from new cures for others. They will die later of causes most associated with very old age. In contrast, the people who clean their hotel rooms, serve their drinks, drive them in taxis to meetings, look after their children, mend their houses and care for them in their old age are unlikely to benefit more than a fraction from the increased longevity that comes from greater overall affluence. This will be blamed on the poor themselves, on their poor lifestyles, on their behaviour, on their claimed lack of aspiration, on their supposed weaknesses. But if it were not for them, what would the affluent have and how would they benefit from their affluence to the advantage of their health? The rich would have to clean their own rooms, make their own food and drive themselves around. They could not rely on others to arrange their holidays, staff their resorts, clean their offices and take their orders. They would have to look after their children when they were very young all day every day and their parents when they were old. The rich could not have jobs where they are paid to travel round the world and could type books about the plight of the poor while other people looked after their basic needs. Most importantly, internationally, the rich countries of the world would not be rich were it not for the efforts of peoples in the poorer nations of the world. We simply could not all consume what the rich consume and we (rich) need the poor to make what we consume. The health of the affluent is as much a product of poverty as is the premature mortality of the poor – worldwide and at home.

Four thousand words, four cigarettes, six cups of coffee, one breakfast and a quick break later and I am finished. In theory, and on average, what I have smoked this morning should reduce my life expectancy by at least an hour. The coffee and the cooked American breakfast won't do much good either. Apart from the quick walk through the lobby I have had hardly any exercise. This hotel appears to have elevators and no stairs. I have been breathing air recycled through a conditioning system. The temperature remains exactly 72 degrees Fahrenheit. Yet I have been doing exactly what I have chosen to do. I have not had to take any instructions from anyone else. I have had no phone calls, received no emails, no students have been able to interrupt my work – I am 4000 miles away from them. I find it hard to imagine a degree of freedom much greater than this. When civil servants in London were monitored to assess the determinants of their health many years ago it was found that those with jobs like mine, which they often described as stressful, but which in reality were not, tended to live longer than civil servants on lower pay grades even when those with more freedoms had what appeared to be worse lifestyles in terms of their health-related behaviour. Had I spent the last few hours doing something directly for somebody else, and that was what I normally did every day, then according to that evidence, the slow cumulative effects of being subservient to others would have done more damage to me than what currently appears to be a very unhealthy few hours spent typing these words. That is all, of course, on aggregate. For any individual anything can happen, and averages are less useful than advice to look after their bodies and their minds. Now, where has that chocolate bar gone...?

8 Work
...the segmentation of society

Whether people work, the work that they do, how well they are rewarded for it and almost every other aspect of employment are, to a large extent, geographically determined in the UK. Furthermore, how those conditions of employment are changing, what work is available, in which industries and the changing extent to which the population is able to carry out this work are also strongly geographically patterned. To understand both how these patterns are changing and what they are changing into requires simultaneously observing processes of both change and concentration. The maps in this chapter are attempts to show both of these distributions together for ten selected aspects of the labour market. By showing this detail it is possible to generalise but also illustrate the degree to which exceptions to such generalisations rule.

People in the UK are segmented, that is partitioned into groups, largely on the basis of the work they do or have done, their parents do or have done and the financial rewards they have received for such work. This segmentation is as much geographical as social and, over time, has become more geographically distinct. There have always been geographical divisions in the labour market, but in the recent past people doing a wide variety of jobs tended to live closer together. Before cars became as ubiquitous as they are today, before many of the new motorways were built, before people had the kind of fear they have of living with others that they have now, there was more geographical mixing of the population and labour market segmentation tended to occur within places between social classes. Now such segmentation is increasingly found between places and within places, there is less differentiation between social classes. The extent to which this generalisation holds varies across the country and depends on how you define a place. London is an exception, for instance, in that in many ways it is an increasingly heterogeneous place. However, if you partition London, as is done in these maps, then that heterogeneity is largely confined to the very centre of the capital.

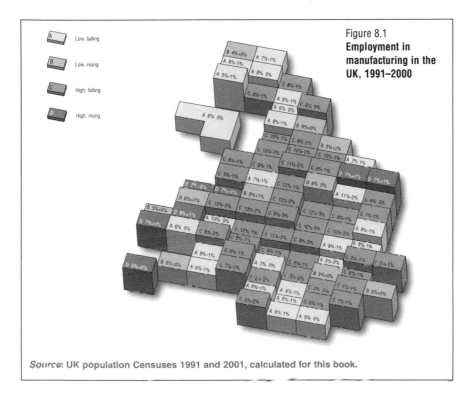

Figure 8.1
Employment in manufacturing in the UK, 1991–2000

Source: UK population Censuses 1991 and 2001, calculated for this book.

Labour markets express the demand for people in places. In some areas that demand is very high, many people move into an area, they tend to be well rewarded and those who cannot contribute with the skills which are required are slowly priced out of the area. In other areas there is very little demand for people. In these places demand was often higher in the past and so there is a surplus of human beings as far as the market for their labour is concerned. People tend to leave these places; the more able leave the fastest. Increasing numbers of people are permanently sick or disabled in such areas, unable to work mostly through the effect that being redundant to the market has had on their health, partly perhaps because this is a rational reaction to the vagaries of markets. Although unemployment, as officially recorded, has fallen everywhere in the last ten years, there has been no significant rise in employment and jobs are now less secure and in many cases more demanding and/or menial than they once were. This may be a major cause of the overall rise in illness rates which is reported at the end of this chapter, a rise which is distinctly geographical in character – as are all aspects of labouring in the UK.

The historical legacy of industry in the UK is etched into our landscape and explains much of our current human geography. Places typified by industries

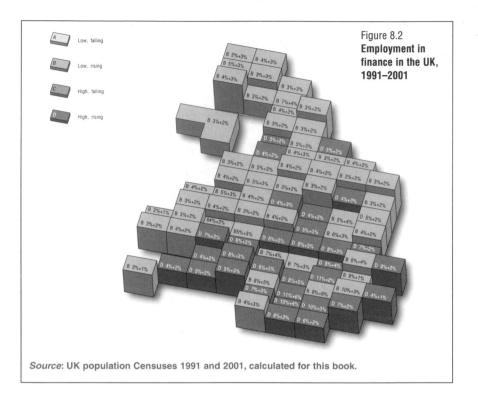

Figure 8.2
Employment in finance in the UK, 1991–2001

Source: UK population Censuses 1991 and 2001, calculated for this book.

in continuous decline tend to appear in relative decline in most other aspects of the lives of the people who live there. Figure 8.1 shows both the proportion of all people working in manufacturing in each area in 1991 and how that proportion changed to 2001. If you sum the two percentages shown in each area, then you will find the proportion working in that industry in 2001. All the figures in this map, and in the nine which follow, are proportions, expressed as percentages, of the entire population to aid comparison. In all these figures areas are shaded according to both whether above or below national average proportions of people were in this category and as to whether that proportion has risen or fallen during the last ten years. Figure 8.1 shows that almost all of the UK was typified by either relatively high rates of employment in manufacturing, which have fallen, or low rates, which have also fallen over time. What the map cannot show is that in the south these rates have fallen because other industries have replaced manufacturing employment whereas that is not the case in the north. The maps that follow are needed to see that. The greatest fall, of the loss of a third of the jobs in this industry in one area, occurred in Birmingham East over the course of the 1990s.

In stark contrast to manufacturing, employment in the banking and finance industries rose throughout the UK during the 1990s, as Figure 8.2 documents. The largest rise was in central London where an additional 8% of the population were employed in this industry in the 1990s, bringing the total in this sector there up from 11% to 19% of the entire population, as compared with 8% nationally. More people now work in this sector nationally than in all of the manufacturing industries combined (which employ 7% of the population of the UK). Work in the finance sector is concentrated in the south east of England and it is there where rises in employment in this sector have been strongest. Distance from London reduces the growth in this sector, which was lowest, nationally, in Cornwall & West Plymouth. Furthermore, as you move away from London the jobs in this sector tend not to be so well rewarded, growth being in areas such as call centre employment. This segmentation can be seen by looking at the changing distribution of occupations rather than industry. For now what matters is that it is the finance industry which has grown the most in the UK over the last ten years. It is ever more concentrated in its historical bastion of London and it is the major source of wealth in the UK. It both supports this country and increasingly divides it. Its profits come from extracting interest payments from the rest of the world.

Figure 8.3 shows that just as employment in the financial industries has risen everywhere in the 1990s, so too has employment in what are termed elementary occupations. These are low-paid jobs deemed by the people who classify jobs to require very little skill to perform, jobs such as cleaning and stacking shelves. These jobs are fairly evenly distributed across the country, but their rise in the south is highest in areas a little distant from London as the people who keep London clean and its shelves stacked with food increasingly have to live further from the capital and have to commute in – a pattern already well established in many North American cities. As the population on average becomes better paid, more skilled, better educated and consumes more, those who can pay increasingly require the services of more and more people in such elementary occupations. Around the coast such work may involve caring for the elderly. In rural areas it might be picking flowers or particular vegetables by hand to ensure the quality which ever more discerning consumers require. In cities it could be the cleaning of the offices which have replaced declining factories, the serving of fast food and child care for the children of women who are now working. Taken together, this disparate group of occupations, all having in common low pay, are on the increase everywhere, but which particular jobs are on the rise depends on where you are looking. The slowest rise has been in central London.

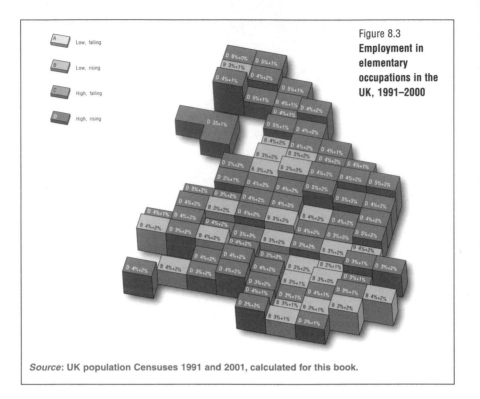

Figure 8.3
Employment in elementary occupations in the UK, 1991–2000

Source: UK population Censuses 1991 and 2001, calculated for this book.

At the other end of the pay scale there have also been universal increases of employment – shown in Figure 8.4 as increasing proportions of the population working in professional occupations (jobs which usually require a university degree and which tend to be well paid). The same proportion of people (4%) worked in professional employment, in 1991, as in elementary occupations, although employment in the former has risen by 1% and in the latter by 2% over the course of the last decade. There are more professional jobs in the south than in the north, especially in and around London. They tend to be better paid in the south, being in the private sector and science, rather than in the state-funded health and education services. The fastest rise in professional employment has been in central London and the slowest in South Yorkshire. Thus not only are professionals better paid in the capital, but there are ever more of them going there whereas there tend to be fewer professionals in the north. In the north their numbers are rising more slowly and they tend to work in professions which, although still generously paid, are not as well rewarded as in the south (when in the private sector). It is almost certainly because there is greater demand for their skills in the south and

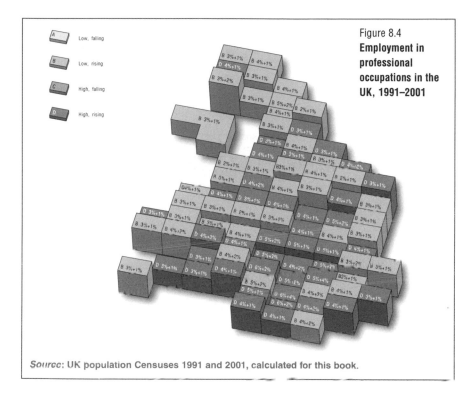

Figure 8.4
Employment in professional occupations in the UK, 1991–2001

A — Low, falling
B — Low, rising
C — High, falling
D — High, rising

Source: UK population Censuses 1991 and 2001, calculated for this book.

hence greater rewards there that this movement is occurring. The source of that increased demand is largely the financial sector.

One way in which demand for labour can be measured is in the number of hours for which people are employed to work. Figure 8.5 shows both the proportion of the population working 31 or more hours a week in 1991 and how that proportion has changed over the course of the 1990s. Although the greatest increases in people working full-time have been in south and central London, the highest proportions are to be found just west and south west of the capital in the most affluent parts of the home counties. Across most of the country the proportion of people working these hours has risen as the demand for labour has grown, but it is in the south where full-time employment remains most common and it is only in East London that a fall is to be found anywhere in the south. Conversely, in Birmingham East, where manufacturing employment fell the most, and in a scattering of other places across the north, the proportion of the population working these hours has fallen as demand for such labour has reduced. The proportion working full-time is also

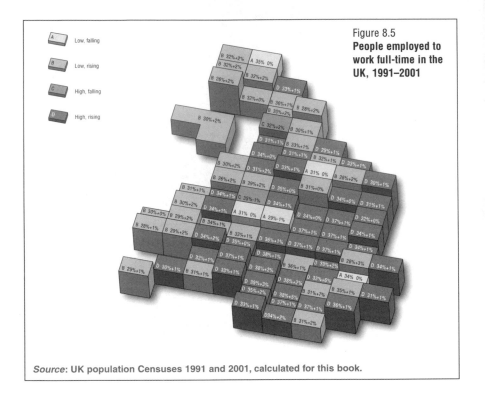

Figure 8.5
People employed to work full-time in the UK, 1991–2001

Legend:
- A — Low, falling
- B — Low, rising
- C — High, falling
- D — High, rising

Source: UK population Censuses 1991 and 2001, calculated for this book.

lowered where large numbers of people are permanently sick (and that has risen), where many are retired and where there are many children (as all the numbers in these maps are shown as proportions of the entire population). In areas with high numbers of children, elderly, or ill people, many working-age adults have to give many hours a week of unpaid time to care for these people. Thus these geographies cannot be understood in isolation.

Geographical divisions by rates of illness present some of the starkest divides in the UK and these are reflected by the geography of mortality shown in Chapter 7. Those people who doubt that such large numbers of people really are as sick as they say they are simply need to compare maps of illness with maps of mortality. One such map of illness is shown in Figure 8.6. This is the map of people who cannot work because they are permanently sick, but are generally of working age. From 1991 to 2001 this proportion of the population rose everywhere, nationally, from 3% to 4% of all people. Outside parts of London and Cornwall, all the areas with above average rates of permanent sickness in 1991 were in the north of the country, as the map shading shows.

HUMAN GEOGRAPHY OF THE UK

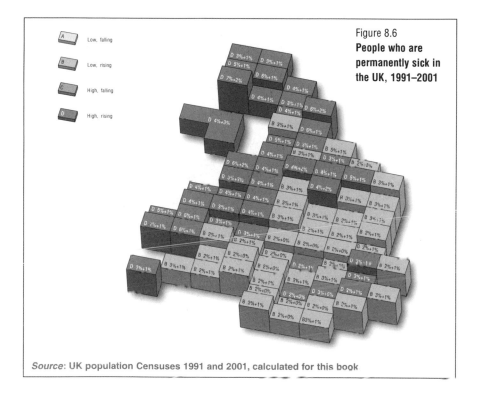

Source: UK population Censuses 1991 and 2001, calculated for this book

The fastest rise has been in Northern Ireland and the slowest in Surrey. It would be difficult to find two places in the United Kingdom which contrast quite so much in so many ways. In Surrey, and most of the areas around it, only 2% of the population are permanently sick and there has been only the slightest of increases to that proportion over time. In Northern Ireland the rate rose from 4% to 7% of the entire population in just ten years and in other largely urban areas of the north it is as high as 8% or 9% of all people now. If these percentages were expressed as proportions of the working-age populations of these places, they would, of course, be much higher.

While rates of sickness have risen universally, if unevenly, rates of unemployment have fallen everywhere in the UK over the course of the last decade. Figure 8.7 shows the inverse of that distribution, the proportion of the population who are not unemployed and how that has changed up to 2001. Children and pensioners cannot be unemployed by Census measures and so everywhere at least 90% of the population fell into this category in 1991 and by 2001 at least 95% of the population of each part of the UK were not

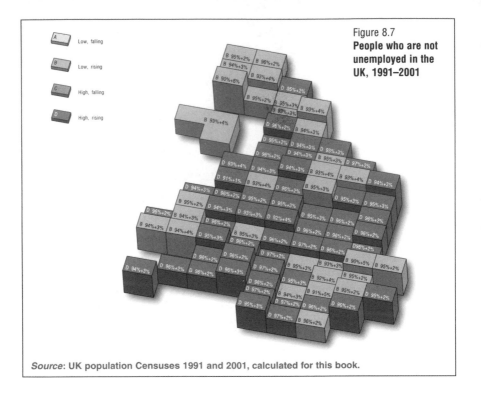

Figure 8.7
People who are not unemployed in the UK, 1991–2001

Source: UK population Censuses 1991 and 2001, calculated for this book.

unemployed. Put another way, maximum unemployment rates, as expressed as a proportion of the entire population have fallen from 10% to 5% during these ten years. However, those changes have not been evenly distributed. The largest fall in unemployment has been in Glasgow, which also had the highest rate in 1991. Glasgow now has the highest proportion of people who are permanently sick in the UK, as Figure 8.6 illustrated. Figure 8.5 shows that it now has the lowest proportion of its population working full-time anywhere in the UK and Figure 8.1 showed that the city had the smallest workforce employed in manufacturing outside central London. Its population employed in the financial industries has risen quickly from 4% to 7% over these ten years, but in Glasgow that means mainly call centre work and, since the Census was taken, such employment has appeared ever more precarious. The quick fall in unemployment in Glasgow masks many other changes.

One change, which has been most extreme in places such as Glasgow, has been in the rise in the numbers of people living in lone-parent families where the parent does not have a job. Figure 8.8 shows that in Glasgow in the 1990s

HUMAN GEOGRAPHY OF THE UK

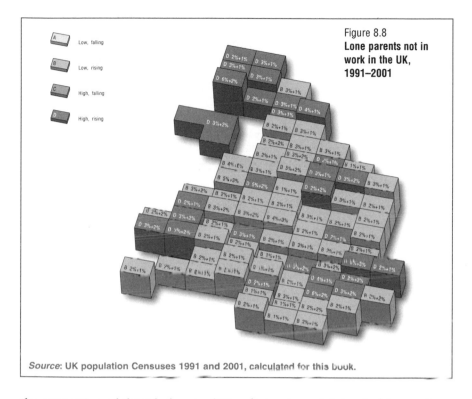

Figure 8.8
Lone parents not in work in the UK, 1991–2001

Legend:
- A — Low, falling
- B — Low, rising
- C — High, falling
- D — High, rising

Source: UK population Censuses 1991 and 2001, calculated for this book.

the proportion of the whole population living in such households rose from 6% to 8% of all people, the majority of these people being children. Thus twice as many people live in lone-parent households with no work as there are unemployed people in Glasgow. Unless you consider the characteristics of the entire population, and all the ways in which they might be economically labelled, you can come to very unrepresentative conclusions by concentrating on a measure such as unemployment alone. Nationally, 4% of all people live in the kinds of household mapped in Figure 8.8, a rise of 1% in ten years. Nowhere has seen a fall in the size of this group. The smallest increase has been in Surrey and the fastest increase in Birmingham East. Note how it tends to be the same places which top and tail the ranking of change by many different measures. It is often impossible to bring up children in Surrey if you are an only parent and have no job. Houses simply cost too much there and there is very little social housing. In contrast, recent industrial decline will have helped to split up more families in Birmingham East in recent years. There is less work for these lone parents and there are more children to be parents of in this area, but housing is cheaper.

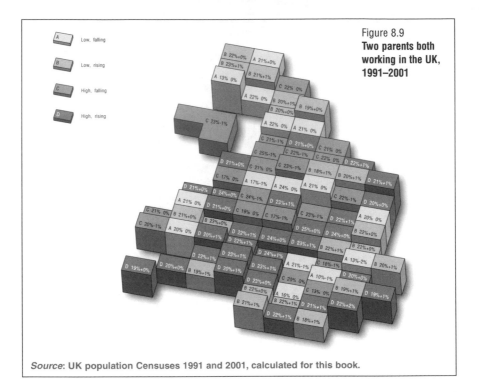

Figure 8.9
**Two parents both
working in the UK,
1991–2001**

A — Low, falling

B — Low, rising

C — High, falling

D — High, rising

Source: UK population Censuses 1991 and 2001, calculated for this book.

In places like Surrey people tend to delay having children until they are older and have established their 'career'. In fact many people cannot afford to move into the area until they have done so. There are fewer children there and so less children to be lone parents of. In contrast, most people who do have children in Surrey tend to be two-parent couples where both parents have to work to be able to afford to live where they do. Thus while high proportions of the children of Birmingham East are deprived of financial resources because the parent they live with does not earn, high proportions of children in Surrey are deprived of time with their parents but not of financial resources. Figure 8.9 shows the complicated map of both the proportions of people living in two working-parent households and the change in that proportion over the course of the 1990s. Outside the major cities, more so in the south, the proportion of people living in such households is high and growing, most so in West Kent. In urban areas the proportion tends to be lower and it is falling most quickly in North East London, principally because there are fewer families with children there. People are now almost two and a half times more likely to be living in this kind of a family in West Kent as compared to

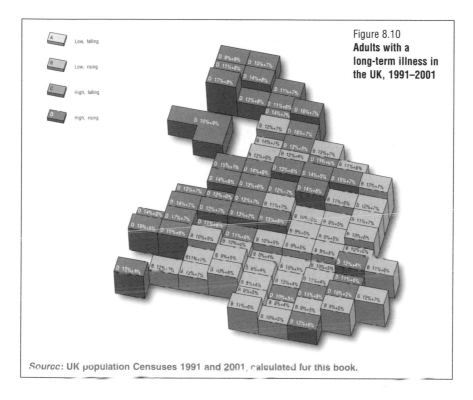

Figure 8.10
Adults with a long-term illness in the UK, 1991–2001

A Low, falling
B Low, rising
C High, falling
D High, rising

Source: UK population Censuses 1991 and 2001, calculated for this book.

North East London. Thus the very nature of everyday family life is becoming more polarised geographically by a combination of economic and demographic changes.

Figure 8.10 shows the most stunning of all the social changes to have occurred in recent years in relation to the segmentation of people by labour. This map is of the proportion of, and changes in, people aged 16 or over suffering from a limiting long-term illness or disability. Nationally this ratio has risen from 12% to 18% of the population in just ten years and that change is not due to ageing – a glance at the patterns to the changes best illustrates that. Both changing demographics and economic circumstances do, however, have a part to play in amplifying the growing divisions seen here. Nowhere are rates of illness falling. They are rising most slowly in the south, and least in London South Inner. They are rising most rapidly in Northern Ireland, but there are also rapidly rising rates in most northern urban areas. In parts of Scotland and Wales over a quarter of the population are adults with a disability and that will quickly be reached in many other areas at current rates of change (to soon be double those rates in parts of the south). Less than a quarter of these people are 'permanently sick and unable to work' (see Figure 8.6). It is thus not an increase in the number of people claiming various benefits which contributes most to this rise. More and more people in the UK are unable to perform tasks, which limits them, due to illness.

An exercise

Segmentation and polarisation are not simple things to measure and there are no set ways of defining them. Take as an example the ten distributions just described in this chapter. In each case the map provides you with both the proportion of people allocated to each particular group in 1991 and how that proportion has changed in the years to 2001, allowing you to also calculate the number of people so allocated at the end of the period. In effect, you have ten sets of data, each set containing two times 85 statistics. Although the numbers are only provided as whole percentages, without having this dataset on computer there are too many distributions to consider for any one person, therefore each select one of the ten distributions to study. If there are 20 or more of you, then you can each also select a year to study, either 1991 or 2001 (although the first and fourth methods defined below allows you to consider both years simultaneously). You need to determine how spatially polarised people were in the year you are looking at for the variable you are considering. There are many ways in which you can do this, some are listed below. Agree a method between yourselves, perhaps more than one, calculate the degree of polarisation which has occurred and then read on. Here are some methods you can use:

1 Most simply, you can say a variable is polarising over time if the majority of areas fall into the shading categories: low and falling or high and rising. However, some variables are rising in every area or falling everywhere. In those cases you can subtract the national average (say median) change from the change measure first, and then redefine every place as either rising or falling in terms of national changes (work out the median change simply by writing down the changes in order and selecting the middle (43rd) one).

2 You can measure the degree of segregation of a group at a given time. For instance, what proportion of people in the country would have to move between areas for that group to be evenly distributed across the United Kingdom? Or what proportion of the group would have to move, or what number of people would have to move, or what number or proportion would have to move if that group were to be distributed as everybody not in that group is! You have a lot of options as to how to calculate just these simple measures of segregation.

3   You can measure the chances of someone chosen at random from the group you are studying meeting another person from their same group if that person were also chosen at random from within their area. This is called an isolation index and is easier to calculate than to describe (it is simply the sums of all the proportions weighted by those proportions, as you can assume each area has equal population). One problem for this index is that it tends to be higher the larger the proportion of a group is nationally. Can you correct for that?

4   You can work out some average changes and draw a histogram of the results. For instance, in areas (grouped) which had a high proportion of people in a group in 1991, what has the average change in their numbers been? What about for average areas, below average areas and so on? For each type of area, as defined by proportion in 1991, there will have been an average change, which you can calculate. You can draw a histogram of those changes. If the tails of the histogram tend to rise and the centre falls, then polarisation has occurred. But what if the pattern is more complex than that?

5   Think up your own way of measuring polarisation or segmentation and change in these measures. Ideally your measure should be simple to understand and preferably simple to calculate. It should measure something which is meaningful, its size should mean something and changes in its size should be readily interpretable. Can you think of a better way of describing whether the patterns shown in the maps above really do represent growing cleavages in the human geography of society or are the changes not as dramatic as that?

Having measured the levels of polarisation or segmentation in whichever way you have chosen to do so, and having looked at the change over time in those measures, you next need to interpret the results. What has been going on? How would you explain your findings to an audience similar to yourselves? Can you appreciate why I have not included such measures here – or would their inclusion have altered the story being told in the chapter above? Should I have included such measures? Having done all this work, what criticisms would you make of this chapter of the book? Can you tell simply from looking at the maps how they are changing? Is the United Kingdom becoming a more or less divided place, at least by these measures? And finally, why consider these measures? What really matters most about places in people's lives?

CONCLUSION

The human geography of the United Kingdom is becoming more segmented by the kinds of work people do, the rewards for that work and the outcomes of those growing spatial divisions. Underlying these changes is the continued decline of traditional industries. Foremost among those, still in terms of the numbers of people involved, is manufacturing. For the first time ever manufacturing now employs fewer people than the financial industries and these shifts have profound geographical outcomes. The United Kingdom has been transformed from an island in which the majority of people farmed or fished, to a country in which manufacturing was the occupation of the majority, to a nation state in which the largest employment sector concerns the movement of other people's monies. The bulk of the jobs, and certainly the best paid jobs in this sector are in the south of England, especially in London. It is thus to London that the most skilled young labour now travels, as we define skills today. It is in and around London that there has been the greatest increase in professional employment, where more and more people are working longer hours, where more children are growing up in households where both adults work, where sickness and unemployment are most rare.

More people are also engaged in the most elementary of occupations as a result of these changes. There is growing demand for people to be employed to do work which in the recent past we would have done for ourselves. We (the rich) now expect and are willing to pay for someone else to make coffee for us in a shop, for there to be millions more metres of shelf space in other shops for us to browse when we choose what to spend our increased earnings on. All those shelves have to be filled. We increasingly expect other people to care for our children in the day when we are at work and to look after our parents when they age (rather than let them live with us). Given the huge rise in menial jobs, greater even than that in professional employment, it is hardly surprising that more people are ill. Add to that the rising numbers who have no job because the industries they used to work in have shut down, and they do not possess the qualifications or youthfulness to work in the new industries, and then add to that the rising numbers of people struggling to bring up children on their own without a wage while others have more money than they need for their children, and it would be disingenuous to suggest that this rise in illness were artifactual. Illness rates have risen most where people are poorest, where rates were highest to begin with and where the majority of people who die prematurely live. This rise in illness has not been translated into a lowering of life expectancies save for in a very few small areas and for a few social groups; it is not a rise in terminal physical maladies therefore. The rising rates of illness are not killing more people although they do provide part of the explanation as to why inequalities in mortality have continued to rise over the course of the 1990s. The patterns shown here suggest that the bulk of the massive rise in illness in the United Kingdom is socially induced: reactions to the growing inequities in life in this country. These inequalities affect people's lives, their abilities to cope with problems, the quality and speed of treatment and the support they receive, depending on where they live for any illness or disability.

Those least affected by their local environments – the young, wealthy and able – are, ironically, those most likely to be able to move to avoid living in an area they perceive as limiting. The same people are most able to change who they are in ways which suit the vagaries of employment. Thus the growth in finance has not been caused by a transfer from manufacturing employment of people who would have otherwise worked in that. It has required migration to occur because the two sectors are not generally located in the same places. Furthermore, they do not generally employ the same people. Many more women work in finance than in manufacturing. A degree is of little use when it comes to actually making things but is apparently essential to being well rewarded in an investment bank or insurance company. The changing industrial geography of the United Kingdom is altering the geographies of occupation, the demographic profile of areas, how families are brought up and where, and overall states of health. And there is no reason to assume that this process is likely to be reversed in the near future. Where there are increases in the proportions of people working in manufacturing it tends to be in places where cottage and small-scale industries are more common. Universities are producing ever more graduates to feed the appetite of the southern labour market. Some graduates will work in call centres, but if they do so they will displace others who could have worked in those jobs and by doing so are not going to reduce inequalities in employment.

Freed from the nineteenth- and early twentieth-century reliance on the physical resources of particular areas for particular industries, buoyant, financially secure and successful companies are more able, in theory, to choose where they locate. But they want to locate within distance of an able workforce, for their workers to be able to travel easily to where they need to go to carry out that work, to be near large international airports and other amenities of a capital, and they want a housing market in which the high salaries they can afford to pay can compete. Given all this, the finance industry is bound to concentrate in the south. Peripheral activities, back offices, call centres, clearing warehouses can be sited outside, but only if they can easily be moved to other locations in the future. In Bristol, Leeds and Edinburgh there are regional offshoots of the major London institutions, but these are simply half-way houses of decentralisation within easy train or plane distance of the capital. They also hardly feature as significant on the national map of this industry. Most importantly, when you can partly choose where to locate your industry, then why place it among the decaying remnants of former industries in areas suffering all the problems of depopulation, illness and worklessness that result? The rise of the car, of working at home, and the repopulation of the centre of the capital have all made more space for the expansion of the finance industries in the south. They have also helped in increasing the segmentation of the labour market and the social polarisation of areas. If there were not a continued and growing demand for unskilled labour – by those who are paid the most – there would be no poor areas in the south of England, especially in London. Exorbitant incomes are of little use if they do not buy you luxuries, and the biggest luxury of all is not having to do things for yourself. Thus even within the south of England we see spatial divisions grow, as they widen too with the north. For them to fall we would have to start living in the same houses as our servants again. Currently we are served in shops and cafés by the tenth of the population who now work there. In the not too distant future it is conceivable that we will begin seeing servants 'living in' again.

9 Home

...the settlements of society

People's lives in the United Kingdom are most influenced day to day by how they are settled and housed – what it is like where they are living. The patterns to housing and settlement are long established yet they are also slowly changing in response to people's changing migration patterns, educational changes, changing social identities, new housing policies/politics, growing economic inequalities, ageing and illness, and the changing industrial and occupational geographies of the country. This chapter begins by considering how closely together or sparsely spaced people are, where major settlements have their greatest demographic influences and how those patterns are changing. Following that we turn to how these changes have altered the types of housing people are living in, then what kinds of household live where – according to their economic position – and how they pay for their housing as a result. The wealth that can be accumulated through housing is our next focus and the attributes of some people's houses and lives which reflect that wealth, where many people live in large houses or with access to many other material goods. Finally, the chapter looks at where most ill and very elderly people live, who has to care for others in their neighbourhood and how people are cared for when friends and family are not able to, or no longer willing to, provide that care.

People in the United Kingdom are housed in many different ways and live in an ever widening variety of different types of household and family, although increasing numbers are also living alone. The types of housing and household people live in are strongly influenced not only by their economic, demographic and social circumstances but also by where they live. In some parts of the country there is very little space per person and even the affluent live cheek by jowl in the densest of city centres (where living alone is an expensive luxury). In other areas, particularly where the population has been declining, there is a general surplus of housing, although few people in such areas can afford to build large homes. Collected together homes form settlements, areas that people have settled, usually for centuries.

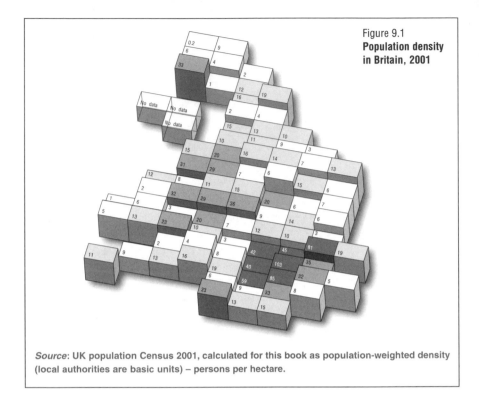

Figure 9.1
Population density in Britain, 2001

Source: UK population Census 2001, calculated for this book as population-weighted density (local authorities are basic units) – persons per hectare.

The structures of these settlements are changing slowly; their nature is changing more quickly. In earlier chapters we considered in more detail the nature of different groups of people living in different areas and how they came to be there, how they are changing the places they live in, how they express their content or discontent through political behaviour, how their lives are economically and socially inequitable, how this translates into patterns of their deaths and how the changing demands for their skills, abilities and time are altering all these pictures. Here we end our tour of the UK by considering how we are looked after through the most basic need we have – shelter – how some are better housed than others and how the pattern of our settlement of this land is changing as a result.

Britain is a very unevenly populated landmass. It is often said to be a crowded island, yet most of the land is only sparsely settled. Figure 9.1 is a map of the average population densities at which people live in each area of the country. This is an average of the densities for each local authority within each of the areas. It better reflects population density as perceived by the inhabitants of each area than does the simple ratio of people to land. The measure varies from

HUMAN GEOGRAPHY OF THE UK

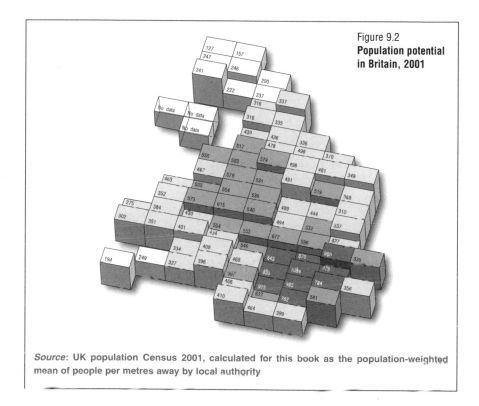

Figure 9.2
Population potential in Britain, 2001

Source: UK population Census 2001, calculated for this book as the population-weighted mean of people per metres away by local authority

over 100 people per hectare in central London (1 person per 10 by 10 metres square), to 0.2 people per hectare in the Highlands and Islands of Scotland (1 person per 250 by 250 metres square). It is often said that in much of Britain people are never more than a few metres away from the rats which live in the sewers. In central London most people are never much more than 10 metres away from the next human being. The population of this island is most concentrated along a diagonal line which runs from Glasgow down through sparsely populated areas along the west coast main railway line, through Manchester to Birmingham, ending in London. They are there (particularly the further north you travel) because that is where most of the housing was built in the past. In England most people in the south live at lower densities than do the population of the north, bar the crowding of the capital.

The diagonal axis of population in Britain is made ever more evident when, instead of considering population density, we consider population potential – how many people are near each person (Figure 9.2). Population potential is simply calculated by summing the population of the country, weighted by the

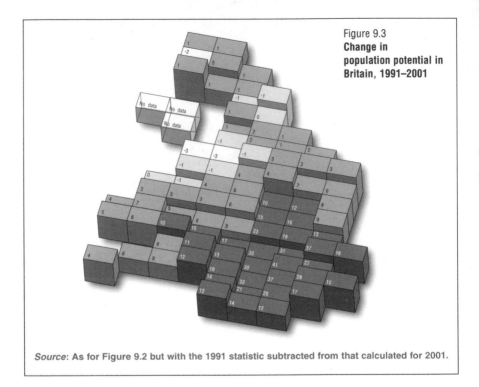

Figure 9.3
Change in population potential in Britain, 1991–2001

Source: As for Figure 9.2 but with the 1991 statistic subtracted from that calculated for 2001.

distance to the place you are calculating it for. It is expressed as the average number of people per metre away and peaks in central London at over 1000 people per metre from the centre of the capital. At its lowest it is just over 100 at the far north of Scotland. Put another way, and given that the population of the country is roughly 58 million, then (very roughly) half the population in Britain lives 580km from the north of Scotland, but only 58km from London. The higher an area's population potential the easier it is for more people to visit there, the greater will be its labour and consumer markets in volume, the higher will be demand for land and space within the area and the population will tend to be more crowded. Had we included the population of Europe or the world in this calculation, and skewed their distance to each place to be calculated as passing through the major airports, then the pattern would be very much more concentrated in the capital than it appears even from this map. London is in the centre. It just appears to be on the edge.

Figure 9.3 plots the change which has occurred in population potential over the course of the 1990s. The changes are of, at most, only a few dozen or so extra people per metre, although what appears to be tiny falls in this measure have implications which are critical for the human geography of the country. The population is rising in the south, centred on a peak of growth in the capital, and is falling in the north west, north east and parts of Scotland. It is those changes which create the very smooth pattern of overall change seen here. As to the importance of these changes to housing: imagine that the supply of housing is largely fixed, a little is built and demolished each year, but there is ever less space to build homes where people most want them (or restrictions on such building which keeps much of the south at low housing densities) and the authorities/owners are reluctant to demolish homes given the cost of building them and the effect such demolition without rebuilding can have on areas. Thus, as the stock of homes is largely fixed, when populations fall their value falls even faster and when there are slight increases in the pressure to live in popular areas this is greatly amplified through rises in the cost of housing in those areas. There are many reasons why the population is moving in this direction. What matters here are the implications of this long-term migratory shift.

Just one example of the effect of population movement can be seen in terms of the changing proportion of people housed in flats, as shown in Figure 9.4. Few flats are built each decade. The population in flats mostly increases because three people live in a flat rather than two, or two rather than one. Similarly, outside Glasgow, which has seen many flats demolished (and hence a 15% fall in the proportion of its residents living in such accommodation), the main reason why the population living in flats is reducing is that where in the past two people lived in a flat, now one does. Thus where the population is falling, increasing proportions of the people who are left live in houses; and where the population is rising, increasing proportions will live in flats. This is one of the simplest examples of how overall settlement change in Britain is altering how people are housed in the country. The rising popularity of living in flats is not simply confined to the capital but can be found across the south east in almost exactly those same places where population potential is rising most quickly. In much of the south east people live at relatively low densities. Thus it is not lack of land which is constraining more and more people to live in flats here. A combination of factors, from the price of housing to needing to be near railway stations to the restrictions placed on new building by the authorities, all play a part.

The type of housing people can afford to live in is as strongly influenced by their ability to pay for it as by what is available to buy or rent, and is being provided by the state and its agencies. Figure 9.5 shows areas of Britain typified by the main economic activity of adults in each area. As with the figures

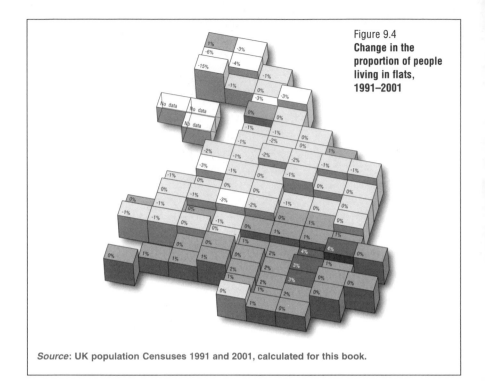

Figure 9.4
Change in the proportion of people living in flats, 1991–2001

Source: UK population Censuses 1991 and 2001, calculated for this book.

shown in Chapter 4, this map is not of what is most usual, but of what group is most overrepresented as compared to the national average distribution of these people in Britain, to make the spatial differences more easily distilled in a single image. Thus much of the north, Wales and Scotland houses more people suffering from a disability than is the norm, and that difference in over-representation is greater for this group than for any other in the area. In Glasgow, for example, there are 6.7% more such people than in the country as a whole and that 6.7% difference is greater than the next largest discrepancy (there being 2.5% more students than average, which cannot be shown by this kind of mapping). In contrast to the north, the south has a large cluster of areas typified by having unusually large proportions of their populations working full-time. Between these extremes there are many other patterns to people's economic status, as shown in this map, but this map is key to understanding how people come to be housed differently across the country.

The past distribution of homes, the supply of land and the economic status of households are key factors in understanding how people come to pay for

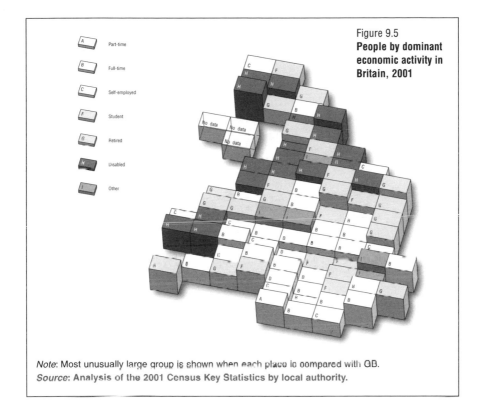

Figure 9.5
People by dominant economic activity in Britain, 2001

Legend:
A — Part-time
B — Full-time
C — Self-employed
F — Student
G — Retired
H — Disabled
I — Other

Note: Most unusually large group is shown when each place is compared with GB.
Source: Analysis of the 2001 Census Key Statistics by local authority.

their housing. But for understanding the map of tenure shown in Figure 9.6, a fourth factor has to be included – demography. A third of the population own their home outright. Most of these people are old and, as the figure shows, they are disproportionately distributed around the coast. To have owned their home, most will have had a mortgage usually initially on a different property and often in a different area. Mortgages are the most common form of tenure in the home counties. Most people who take out a mortgage rent privately before doing so and renting is the typical atypical (i.e. unusually large) tenure of much of London. Private renting is, however, the most expensive form of tenure in the long run and many people cannot afford it. Much private rented stock is also only suitable for young childless adults. A fifth of all households are housed by local councils or housing associations where the state or its agents are the landlord. This tenure is most typical of the north of England and Scotland, although is also overrepresented in three areas of London. Compare this map of how people pay for or own their homes against that of what people do in the economy, the maps of age structure

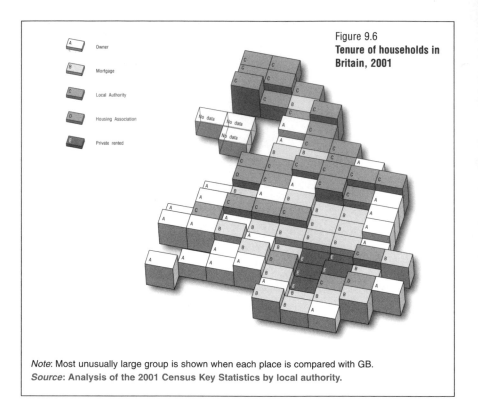

Figure 9.6
Tenure of households in Britain, 2001

A — Owner
B — Mortgage
C — Local Authority
D — Housing Association
E — Private rented

Note: Most unusually large group is shown when each place is compared with GB.
Source: Analysis of the 2001 Census Key Statistics by local authority.

and other identities shown in Chapter 4, of how people vote (in Chapter 5), of what they earn (in Chapter 6), and of how they die (in Chapter 7).

Homes, of course, differ not only in how they are paid for and how they are built, but in what you get for your money and your land. Figure 9.7 shows the proportions of people who live in homes with seven or more main rooms in them. These are generally the most expensive properties in each area, but in some areas almost a third of all homes are as large as this. Where people already have a kitchen, dining room and sitting room on the ground floor and three bedrooms above them, the addition of a single extra bedroom through building an extension will bring a home into this category (toilets and other small rooms are not counted here). Seven-room houses are most rare in Glasgow, despite the recent demolition of so many flats there. Next it is within London that fewest are found, despite the immense amount of money flowing through the capital. Where there is either lack of money or lack of space, such housing is rare. It is most commonly found in the south,

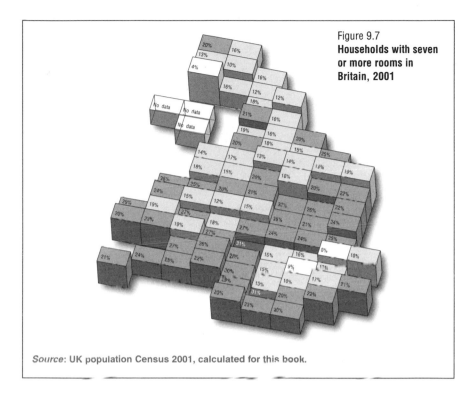

Figure 9.7
Households with seven or more rooms in Britain, 2001

Source: UK population Census 2001, calculated for this book.

although the proportions of large properties to be found in Wales are high. However, a large proportion of those will be owned by people who have retired from England. People who are retired need no longer live within commuting distances of major workplaces. Moving towards where land and properties are cheaper allows some to buy larger and more comfortable homes than they could have lived in when working in England. It also results in some people working in Wales having to live in smaller homes than they otherwise would.

The tenure of housing, its type, worth and how large it is tells us some aspects of how people are housed but not much about what is actually within those walls – what possessions people own or have access to in their household. Many surveys are made of these but few result in information which can easily be mapped, even at the crude level used in this book. One thing which is consistently asked in population censuses and which does have a great influence on housing is how many cars, if any, each household has access to.

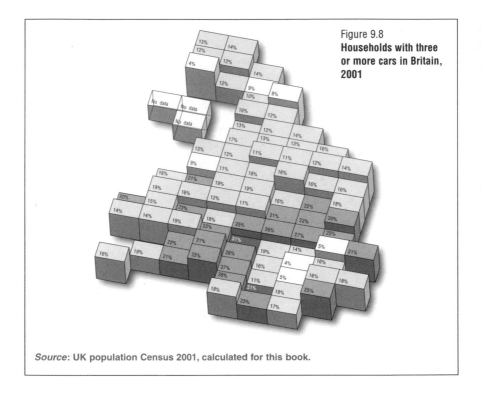

Figure 9.8
Households with three or more cars in Britain, 2001

Source: UK population Census 2001, calculated for this book.

Figure 9.8 shows the proportion of households which have access to three or more cars. It is a strikingly similar distribution to that of rooms, albeit with fewer cars in retirement areas where more households consist of just one or two people, some too infirm to drive. The two areas of the country where 31% of people lived in homes with seven or more rooms also see 31% of their households having access to three or more cars. It is partly through owning so many cars that affluent middle-aged people have managed such a take-over of the home counties, which only a few decades ago were rural enclaves rather than commuter settlements. They also have space to house the cars as well as the money to afford to buy and run them. In the cities there is simply not the parking space on roads for so many cars per household; and there are other forms of transport available, especially in London, and so we see a clear urban/suburban divide.

Car ownership tends to be lower in the north even where there is space to park vehicles and in rural areas where they might be considered more of a necessity. People in the north spend less time in cars partly because they cannot

HUMAN GEOGRAPHY OF THE UK

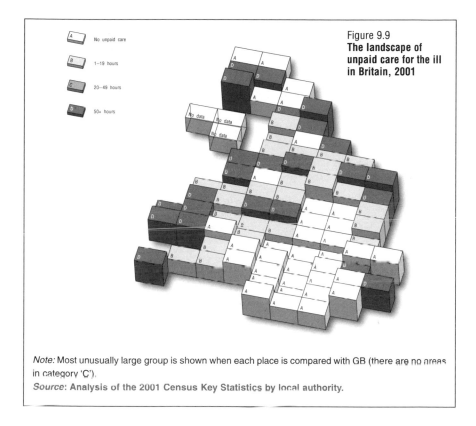

Figure 9.9
The landscape of unpaid care for the ill in Britain, 2001

A — No unpaid care
B — 1–19 hours
C — 20–49 hours
D — 50+ hours

Note: Most unusually large group is shown when each place is compared with GB (there are no areas in category 'C').
Source: Analysis of the 2001 Census Key Statistics by local authority.

afford to drive so much. They also have other calls on their time. Figure 9.9 shows how it is in the north that people are most likely to have to give up 50 or more hours a week caring for a relative or neighbour who is ill or infirm, while the south is dominated by areas where what is most disproportionately high are the numbers of people who do not need to provide any such care. Although rich and poor can be found everywhere, there are increasingly two Britains. In one, at the extreme, you will find homes within which two adults work and are well rewarded, where everyone has a bedroom to themselves and even the teenage child has their own car, where mum and dad have almost paid off the mortgage and have seen the value of their property double several times since they first moved in. In contrast, in the north, and at the other extreme, you will find smaller homes where a single elderly and infirm person lives, whose daughter visits often to look after them. This person has no car and no capital as they pay their rent to the council, and when their neighbours die occasionally no one new moves in.

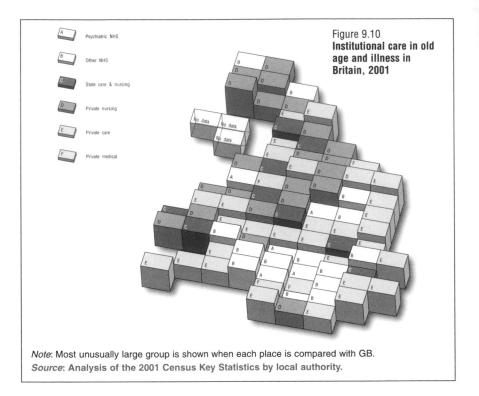

Figure 9.10
Institutional care in old age and illness in Britain, 2001

Legend:
A — Psychiatric NHS
B — Other NHS
C — State care & nursing
D — Private nursing
E — Private care
F — Private medical

Note: Most unusually large group is shown when each place is compared with GB.
Source: Analysis of the 2001 Census Key Statistics by local authority.

Figure 9.10, the final map of this chapter, is one possible depiction of where most people end up being housed towards the end of their lives and how that currently differs across the country. The map is of the population who live in the institutions (hospitals or various 'homes') listed. The most atypically populous type of accommodation is shown, rather than the most common, to highlight the geographical patterns. London and much of the non-coastal south is unusual in that here relatively high numbers of people are found in NHS or private hospitals. These show up here mainly because the other institutional populations are so low in these areas. In practice hospitals mostly cater for a population which moves in and out of them rapidly. From the Welsh border all around the southern coast to Lincolnshire an almost uninterrupted swathe of private care homes dominate. In the north, Wales and Scotland it is private nursing homes that are more typical. Often the authorities pay for people to be housed there when they can no longer cope at home and need medical care. Care homes are more of a luxury, often for the children of the elderly (who then need not care for their parents). There are far fewer private care homes in the north.

An exercise

Try to imagine how the country might look if this book were to be rewritten in 50 years' time. Today's school leavers will be of pensionable age, but which will have good pensions? Where will they have moved to and how will they be accommodated? Many of their parents will still be alive if life expectancy continues to rise as it has done for the last 50 years, but who will be caring for them? There will be fewer people of working age and fewer children again, unless today's school leavers behave differently from their parents or unless more young people come into the country than leave it. What will the housing stock be made up of? How many of the Victorian terraces will still be standing? What will be the state of the homes built around the middle of the last century, now all at least a century old? And will some families still have the state or its agents as their landlords? Which settlements will have declined and which will have grown? Will the old still move to the coast? Will the young still cluster in university towns and move in large numbers to the capital? Will fewer or more people be ill? Of what will the population now be dying? What could the human geography of the United Kingdom look like in, say, 2055, if there is a 'united' Kingdom then?

Speculation over the future is a fraught but interesting exercise. One way to conduct it is to divide your speculation up by how uncertain you are about different issues. Start with issues you think are more certain, move on to things which are more uncertain, and end with pure speculation. To start you off here is one possible way in which you could begin with a set of issues to consider:

1   More certain: We largely rely now on infrastructure built over 50 years ago – roads, rail, sewers and houses. Thus we largely live in the same places we lived in half a century ago. What is the state of this infrastructure? What kinds of things could you expect to see built in the coming years? Airports often take decades to plan and build for example. And what might the impact of such changes be?

2   Less certain: Some aspects of human life change slowly and in one direction for long periods of time, for example the fall in fertility, the rise in life expectancy, people being educated for longer and longer periods, national wealth rising, inequalities not tending to diminish. If those aspects of life in Britain which have changed slowly and in a steady direction for most of the last 50 years were to continue on their way, what would the future hold for us? What if you include trends which appear to be cyclical such as economic recessions?

3   Pure speculation: Most forecasts of the future appear to work until something unusual occurs and there are many unusual things which can occur. The human geography of Britain was last altered significantly by a major infectious disease pandemic and war over 80 years ago. How would we cope with such an event in the future? We were last on the receiving end of a major war over 50 years ago. War has far from ended around

the globe, so what if we were at war at home again? Both 100 and 50 years ago radical governments took power in Britain and instigated many changes to improve people's lives. Could that happen again? And what could happen that you have not thought of?

Having thought up a set of issues and characterised them into the three groups above, set to work on outlining possible scenarios for the future, and divide the work up. Draw possible maps of the future. These are far easier to draw than are maps of the present because you can simply make the data up. All you need is for your map to be plausible. For instance, draw a map of the results of a fictional general election in 2055. What kinds of political party might there be and what voting system? Will some children now be allowed to vote or will voting rights be more restricted? You may think it unlikely that there will be such elections in 50 years time, after all, elections in which almost all adults are allowed to vote are less than a century old in this country and there have been only roughly two dozen of them, hardly a long time series. But even if you think like this, try to draw a map of how the alternative to elections might operate.

The one thing you can be (quite) sure of is that there will still be a distinct geographical pattern to the lives of people on and around this island. There always has been. Perhaps the hardest future to imagine is one in which every place were the same as far as the people living there were concerned. You could move around the country but you could not tell where you were from, what the people there were doing, how they were living and what they had. It is a difficult future to imagine unless, that is, you were a child of the late 1970s and early 1980s in this country. Then, if you believe the rhetoric of government, there would be equality in the future. The poor would grow rich on the trickle down of monies from the affluent, state housing would all be sold to its inhabitants and the future was a rosy, prosperous hard-working utopia for all, even for the inhabitants of 'those inner cities' for which they had 'task forces'. If you believed some of the opposition of those days, then the future was equally equitable if a little more bleak. With America's aid we would manage to engage in a nuclear war. Most of us would be killed and the survivors would 'envy the dead'. There would be little variation in the remaining conditions of living across the Kingdom (which would almost certainly no longer be a kingdom), although you'd be best placed to try to get to the Caledonian canal which would have become the major new trade route for a rapidly emerging new stone age.

Safe to say neither of those predictions came true. The opposition came to power and it now says that we will live in equality in the future as they 'bring Britain back together again', but few believe they will. If you believe other voices, including the barely concealed voice of parts of the government, then what we have most to fear is a bomb or a virus, now not from the Russians, but still hitting our major cities. There is always speculation and in some ways it doesn't change a great deal – the authors and actors are altered but the scenes portrayed remain much the same. Unless I am extremely lucky (or unlucky), I won't be around to see what the world looks like in 2055. For my generation and older, your predictions cannot be proved wrong.

CONCLUSION

How we are housed, from when we are born through to where we die produces patterns on the maps of life in Britain which reflect a great amount about us and how we have settled this country. In the UK almost everyone is given shelter, but the shelter they receive varies greatly and almost certainly more across the country than it has ever done. To the north are the old industrial cities where the stock of small, often nineteenth-century, housing is ageing. To the south many more homes were built in the last century and far more of these have been enlarged and modernised than in the north. For over a century the population of Britain has been moving southwards. In recent years large numbers of people have begun moving (net) into London again (which for many decades before had been becoming less crowded). However, those now moving to London are bringing money or qualifications which will gain them future wealth – but no amount of money can create more space. Thus in the capital even the very rich have to squeeze into smaller and smaller spaces. In the last ten years the numbers of people having access to many cars in the capital has declined. Saturation point for the parking of vehicles was reached while the population continued to grow. There are thus fewer cars per person in the capital now than there were a decade ago. In contrast, around London the population is still relatively sparsely distributed. Strict planning controls, very high land prices and problems of commuting mean that populations have grown far more slowly around the capital as compared to within it. However, those who can live in the home counties have quietly added rooms and wings to their properties, increasing their value further. In many places there are almost as many cars here as adults – another saturation point has been reached as there are so few left who could drive. People here are finishing off purchasing their houses and further out the majority have done so. This is where the housing wealth of the United Kingdom is accumulating most clearly. There are exceptions to these generalisations, pockets of poverty in the south, but outside the poorest parts of London such pockets are shrinking and mile upon mile of land is becoming much the same in terms of who can now live there. Around the coast of this half of the country are increasingly found the parents of the people who have just paid off their mortgages. These pensioners are living in care homes which are often paid for from the proceeds of their own house sales in the past. Their grandchildren will only be able to buy where they once lived if they help them. In contrast, in the north, this cycle of money begetting money, of house prices ever rising, of ever increasing demand to settle is far more rarely found. In parts of the north small areas are being abandoned and larger places depopulated. In many areas there is an elderly population living in flats for which there are too few people to replace them when they leave. In general they will not leave until the state pays for them to enter a nursing home or they die. It is also in the north, in this kind of a housing market, that most people too sick to work are found, where the state and its agents still own large amounts of local housing, where there are far fewer large homes and families with more vehicles than they need. It is in the north too that the highest numbers of people need to devote the longest time to the unpaid care of others. The very last of the dark satanic mills have finally closed, but the green and pleasant land of Britain is only for the few who can afford it.

10 Abroad

...the Kingdom's place in the world

From speculating on the future, to trying to understand the present and recent past of the human geography of the United Kingdom, a wider view of the world is required. It is unlikely that the future maps of the human geography of this country will be determined from within, as neither the current nor the past maps have been. The world outside the UK has been ignored in these pages so far, but its influence is clear to see through almost every figure, pattern and event described. The basic map used in these pages is of areas which were created for the election of members of the European Parliament (Figure 1.1) and yet almost nothing has been said on the rest of Europe. The landscape underlying the map is of children's chances of entering university (Figure 1.5). The main reason those chances are historically high is that British governments have been trying to catch up with American and other western enrolment rates, the rush to educate was not internally inspired (Figure 1.6). The numbers of people living in this country and the variation in their ages are products of world wars and other international events (Figure 2.1). Our graduates increasingly flock to London to directly or indirectly serve international financial institutions (Figure 3.10), as do many people from the rest of the world (Figure 4.8). For those trying to understand why more and more of us do not bother to vote (Figure 5.6), the rising international disillusionment with such participation might help explain that such trends are not purely home grown, neither are the increasing variations in our incomes (Figure 6.2), as recorded by international banks which 'invest' our money around the globe. We can easily see how rates of some diseases depend on external factors (Figure 7.2), but it is international markets that move people around Britain over decades, rewarding some areas and depleting others of resources that result in such stark geographical divides in the ways most of us die (compare Figures 8.2 and 7.7), and which is slowly altering our basic pattern of settlement in this land (Figure 9.3). The rest of the world has a far greater impact on us than we now have on it. It is from where we make our profits to generally live a good life, from where much of our food and most of what else we consume comes, it is often because of economic competition with the rest of the world that our politicians tell us we have to work harder and longer and why they cannot afford to remedy inequalities within our borders. However, social inequalities within our borders are as nothing compared to those now seen across the globe. Both our distant imperial past and recent world banker re-emergence have helped to foster these divisions. In this chapter the story of the human geography of the United Kingdom is brought to a close by looking at where that Kingdom now sits in the world. Because this book began by looking at the life chances of children in Britain, it ends by looking at the chances of children across the globe, and at how the UK sits within a far greater map of human geography. This is partly a map of the future as it is these children who will determine the fate of the children of our islands in the long term. How does the world look when drawn to depict the lives and chances of its children? How does the Kingdom fit within that map? What is abroad?

The world as viewed by its children is a world dominated by Africa, India and China. The United States and Russia are relatively insignificant. The United Kingdom is separated by oceans and rich nations from the immediate lives of most of the world's children. Its child population is less than half the size of Egypt, Ethiopia or the Philippines. It is less than a quarter the size of Nigeria, Brazil or Bangladesh, less than a fifth the size of Pakistan or Indonesia, and there are almost 30 children in both China and India for each child living in the UK. This is the current map, the future map sees these disparities in population rapidly widening. Social inequalities in life chances, in access to education, in work, and in health within the UK pale into insignificance when compared with such inequities across the globe. However, much the same forces which divide up children's chances geographically within one affluent country are at play worldwide. People are of different monetary value depending on where and to whom they were born both within Britain and within the world. Where there are more people than markets appear to require, the results can be dire and increasing numbers of people rapidly move around the world to avoid such consequences. As that movement continues and as the lives of these children are played out in this century, what happens to the world's children will determine what happens to the children of Britain in not too many years to come. Up to this point in these pages it is as if we have been speculating on the state and future of a rich neighbourhood while ignoring the city in which it sits. If the world were a city, the UK would be an affluent neighbourhood offset from the core of that city, but ever less immune from its future, its rapid growth, its claims for fairer resources and its abject poverty.

What most clearly divides the world's children is the poverty many are growing up in as compared to the affluence of most of those living in the rich nations of the world. In 2003 a report was published that sought to 'produce the first accurate and reliable measure of the extent and severity of child poverty in the developing world using internationally agreed definitions of poverty' (Gordon, D. et al. (2003) *Child Poverty in the Developing World*, Townsend Centre of International Poverty Research Report for UNICEF, University of Bristol, UK, p. 7). In this chapter some of the data presented in that report is used to map many aspects of poverty around the world. The report used data from demographic and health surveys in 46 countries that contain over half of the world's children to calculate the numbers of children living in poverty in each place. Here the data for all 45 of those countries which each contained more than half a million children are depicted. Eight of those countries are named in Figure 10.1 (where the UK and USA are also identified) and the further 37 are listed in Figure 10.1 which provides a key to the countries. In these surveys households containing in total 1.2 million children were interviewed in recent years. From the results of those surveys, in each (except one: Figure 10.7) of nine maps which follow, these 45 countries are shaded light grey if less than a third of children are suffering from the form of poverty being depicted, medium grey if the proportion is between a third and two-thirds of all children and dark grey if over two-thirds of all children in these countries are so affected. The rest of the world is shaded white because no comparable data were available for the children of those countries. In the affluent countries of the world almost no children will be suffering the forms of poverty as detailed in these next nine pages. The final two world maps in this series chart two different measures of the aggregate burden of poverty on the lives of the majority of the children of the world.

For children, severe water deprivation is defined as only having access to surface water (i.e. not piped) for drinking, or living in a household where the nearest source of water was more than 15 minutes travel time away. Such children are deprived both of water by quantity and almost certainly through its quality as well. Among the countries plotted in Figure 10.2, over half of all the children living in Rwanda, Uganda, Ethiopia, Madagascar, Tanzania, Kenya, Cambodia, Mozambique, Chad, Cameroon, Malawi, the Central African Republic and Ghana are deprived of water in this way. All of these countries were governed by colonial European powers in the past, including the British.

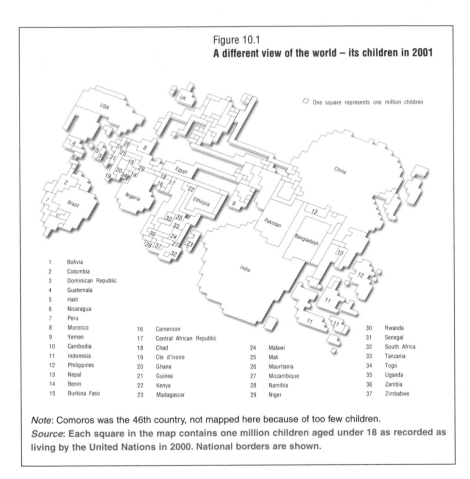

Figure 10.1
A different view of the world – its children in 2001

One square represents one million children

1	Bolivia
2	Colombia
3	Dominican Republic
4	Guatemala
5	Haiti
6	Nicaragua
7	Peru
8	Morocco
9	Yemen
10	Cambodia
11	Indonesia
12	Philippines
13	Nepal
14	Benin
15	Burkina Faso
16	Cameroon
17	Central African Republic
18	Chad
19	Cte d'Ivoire
20	Ghana
21	Guinea
22	Kenya
23	Madagascar
24	Malawi
25	Mali
26	Mauritania
27	Mozambique
28	Namibia
29	Niger
30	Rwanda
31	Senegal
32	South Africa
33	Tanzania
34	Togo
35	Uganda
36	Zambia
37	Zimbabwe

Note: Comoros was the 46th country, not mapped here because of too few children.
Source: Each square in the map contains one million children aged under 18 as recorded as living by the United Nations in 2000. National borders are shown.

Over a fifth of all children in the poor majority of the world only have access to unsafe or distant sources of water, and they are most likely to live in these conditions if their land was once directly governed by the countries in which such deprivation is now unknown. It is the children living outside the towns and cities in these countries who are most likely to suffer such deprivation, although in cities piped supplies can easily be contaminated with sewage through leaks. The children of the world who are deprived of safe water are, of course, much more likely to die in childhood or young adulthood, but billions will survive for longer. Diseases flourish where water supplies are poor and human beings are weakened through these deprivations. A century and a half ago cholera epidemics were commonplace in London due to the quality of the water. For those children who survive these deprivations today, a tiny minority will later travel to affluent nations and see for themselves the riches

HUMAN GEOGRAPHY OF THE UK

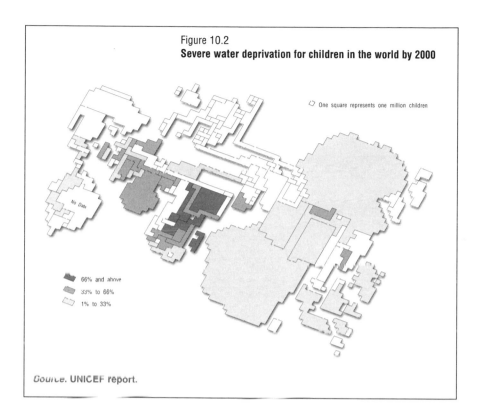

Figure 10.2
Severe water deprivation for children in the world by 2000

One square represents one million children

No Data

66% and above
33% to 66%
1% to 33%

Source. UNICEF report.

there. The monies required to supply every child in the world with clean water are far less than the countries of the west spend on their pet food.

It is not a lack of worldwide resources which deprives children of water, but lack of political will to provide it. The children who do not have access to safe water are those children whom people with power do not value. However, depending on where they live, the children of the poor majority of the world have differing chances of being deprived of different things. The particular histories of countries, their physical geography and current economic worth all influence these patterns. Figure 10.3 shows where the children most deprived of sanitation live. These are children who have no access to toilet facilities of any kind in (or near to) where they live. For these children there are neither private nor communal toilets nor even pit latrines, nor is there any systematic means of sewage disposal. Over half the children living in Nepal, Ethiopia, Cambodia, Niger, Burkina Faso, Benin, Chad, India, Namibia, Togo, Madagascar, Mozambique, Yemen, Mauritania and Pakistan lack such facilities. The largest group live within what were the borders of

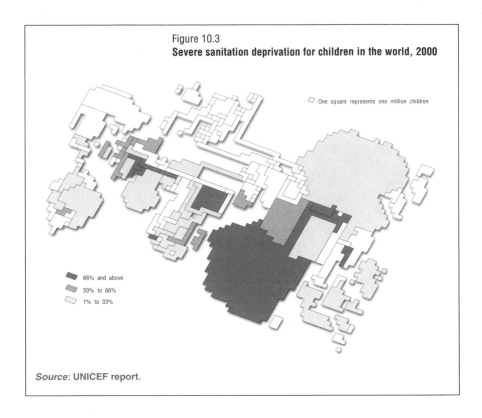

Figure 10.3
Severe sanitation deprivation for children in the world, 2000

One square represents one million children

66% and above
33% to 66%
1% to 33%

Source: UNICEF report.

India, as defined when the British ruled the subcontinent. Britain was among the first countries in the world to introduce near nationwide sewerage and sanitation infrastructure which helped end the cholera outbreaks in that country and turn the tide of many other diseases. In fact it has been claimed that the introduction of this infrastructure did more to improve the health of the population than any medical intervention. To afford such facilities we needed the profits gained by owning India. There is thus a direct link between the sanitation that children in the United Kingdom enjoy today, much of their sewage still flowing through Victorian sewers, and the lack of sanitation seen across India and in other parts of the globe which were once governed from abroad to extract the profits that made the affluent countries rich.

Shelter is the most basic of necessities in the world. Children can travel to find water, can use open ground for sanitation, but without adequate shelter they will physically suffer regardless of their efforts. Being severely deprived of shelter here is measured as living in a place where there are more than five

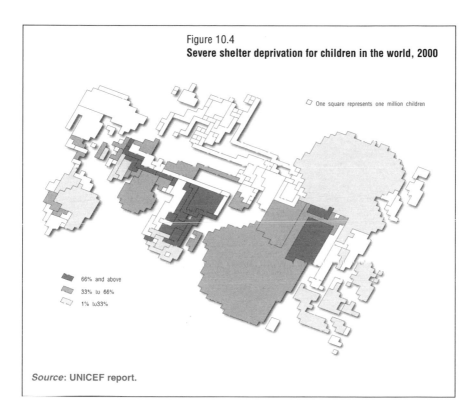

Figure 10.4
Severe shelter deprivation for children in the world, 2000

◻ One square represents one million children

66% and above
33% to 66%
1% to 33%

Source: UNICEF report.

people per room or which has mud flooring. Some 34% of the children of these countries live in such circumstances (compared with 31% who are most severely deprived of sanitation). Such accommodation provides near perfect circumstances to harbour disease, to never have privacy, and to spread infection. Because the data on which these figures are based are derived from household surveys, they provide no measures of children who are completely homeless. A majority of children in the following countries are severely deprived of adequate shelter: Chad, Ethiopia, Nepal, Bangladesh, Rwanda, Uganda, Niger, Malawi, Tanzania, Central African Republic, Mali, Mauritania, Burkina Faso, Mozambique, Kenya, Namibia, Nicaragua, Zambia, Yemen, Guatemala, Cameroon, Guinea and Peru. The countries in this list (and the other lists in this chapter) are sorted in order from 96% of all children in Chad, to those where the rate is nearest 50%. Compare Figure 10.4 with the map of people living in homes with seven or more rooms in Britain (Figure 9.7), and try to imagine living with six, seven, eight or more people in a room, or with a floor made of mud. Note also how rates are highest in African and India, and lower

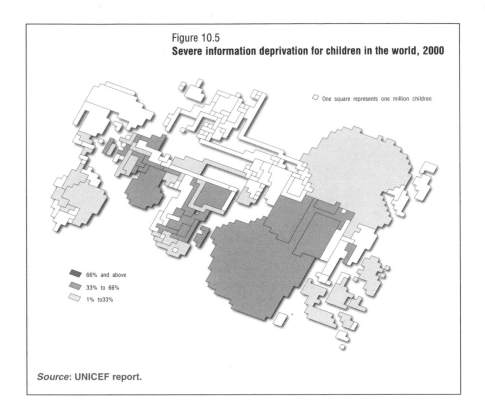

Figure 10.5
Severe information deprivation for children in the world, 2000

◇ One square represents one million children

■ 66% and above
▨ 33% to 66%
▱ 1% to 33%

Source: UNICEF report.

in countries usually not governed, at least directly, by colonial powers in the past, such as in South America, parts of East Asia and China. In fact the lake around which the greatest concentration of children in the world live without adequate shelter, shown on the map in Africa, is named after Queen Victoria. Her name is also used to describe a form of spacious housing which began to be built in the UK during her reign.

In the poorer countries of the world children are not simply deprived of the physical necessities of life – shelter, water, sanitation and (as we turn to shortly) food. They are also deprived of education, health care and basic information. Severe information deprivation is defined as the proportion of children aged between three and 18 who have no access to newspapers, radio or a television where they live. Globally, a quarter of all children in the poor majority of the world are deprived of such basic sources of information. Again these children can be seen from Figure 10.5 to clearly be most concentrated in the majority of Africa and the Indian subcontinent. However information deprivation is only

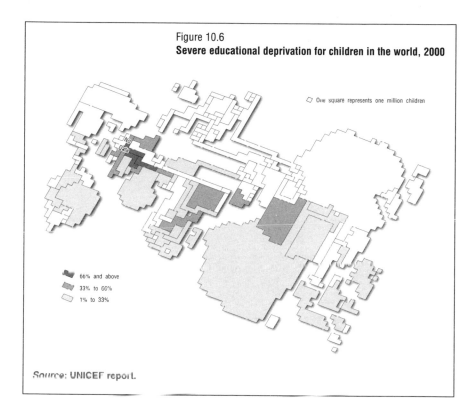

Figure 10.6
Severe educational deprivation for children in the world, 2000

One square represents one million children

66% and above
33% to 66%
1% to 33%

Source: UNICEF report.

the experience of a majority of children in Benin, Ethiopia and Chad. Elsewhere most children do have access to one of these forms of information, more often than they do to clean water, shelter or sanitation. The very fact that such a high proportion of children do have access to such information illustrates the degree to which information is valued. A majority of the world's children are illiterate to a degree. Even in areas of Britain up to almost a quarter of adults are judged to be functionally illiterate (Figure 6.7). Thus radio and television are the primary means by which information can be transferred and both these means require a broadcast network and a power grid if all but the most basic of radio signals are to be transmitted and received. It is where both education levels are lowest and such infrastructure is least developed that information is most scarce. As information without knowledge is of less value, so it is to education that we turn next.

Figure 10.6 maps the proportions of children in the 45 countries being studied who are severely deprived of education. This is the proportion of children in each country aged between seven and 18 who have never been to school. They are thus children who have received no formal education of any kind. Partly because the proportion of children in this situation in China is less than 1%, the international proportion is only 13% of all children. However, this is a very strict definition of educational deprivation. Spend one day in a school in your life and you are no longer deprived by this measure. Over half the children of Niger, Mali, Burkina Faso, Ethiopia, Chad and Guinea are severely educationally deprived even by this strictest of standards. Girls are far more likely than boys to be so deprived and deprivation is almost everywhere greater in rural areas. Such a pattern would not be out of place in Britain two centuries ago and puts the current maps in Chapter 3 of this book in stark context. While the underlying landscape of the United Kingdom depicted here has had its topography determined by a child's chances of attending university, the world map of educational disadvantage concerns whether a single day of the most basic primary education has ever been received. Such a measure hides far greater proportions of children who receive only the most limited education and never learn to read or write at a simple level, and it disguises the impossibility of ever attending a university for the majority of the world's children. The measure does, however, highlight where those who receive the least education are most likely to be living today.

Some 15% of the children of the poor majority of the world are severely deprived of food. The definition of this deprivation is that it includes all children who are so severely malnourished, stunted and emaciated that their heights and weights are more than three standard deviations below the median of the international reference population. Thus the measure includes all children who are most severely wasted, stunted or underweight. To allow for international comparability the proportions reported here are for all children aged under five years, but they are representative of older children too. In no country are as many as half of all children routinely so severely food deprived. The highest proportions of between a quarter and a third of all children being so severely undernourished are found in Bangladesh, Niger, Ethiopia, Nepal, India, Mali and Madagascar. Thus Figure 10.7 would simply depict a single shade for all those countries had the key not been altered. Rates were lowest in Colombia, the Dominican Republic and Brazil. Although this is a very severe measure of food deprivation, it is useful to consider that the researchers who decided how to grade each form of deprivation used in the UNICEF report concluded, in effect, that lack of access to clean water, sanitation and shelter are more widespread than continued severe lack of access to food. The stereotype of the children of poor nations starving is misguided. Starvation on a greater scale than this would result in population declines. Note again,

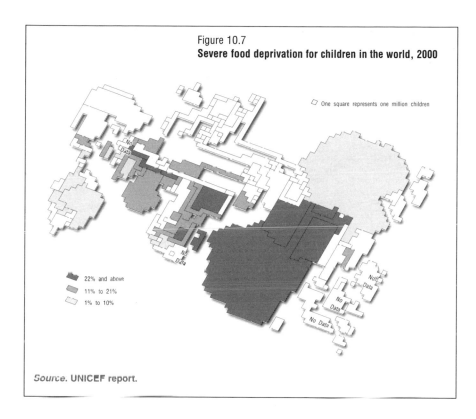

Figure 10.7
Severe food deprivation for children in the world, 2000

One square represents one million children

No Data

No Data

22% and above
11% to 21%
1% to 10%

No Data

No Data

No Data

Source. UNICEF report.

however, that it is in many of the places which Britain once governed where the most children go continuously hungry to an extent that this severely deforms their bodies by age five. There is, of course, far more food to be eaten than these children could ever consume, and in the United Kingdom some children are consuming far too much. The problem is not its production, but distribution, which is currently done by how these children are valued.

The rarity of widespread severe starvation is one reason why the population of children in the poorer parts of the world can still rise rapidly. The other major reason is health care and in particular inoculation. For Figure 10.8 severe health deprivation is defined as a child who has not been immunised against any diseases and young children who have had a recent illness involving diarrhoea and have not received any medical advice or treatment. Diarrhoea is the most common cause of death in infants in the world, although it is such a rare cause of death in Britain that it is not separately identified among the hundred causes of death which end Chapter 7. (In Table 7.1 it is a component of the first cause,

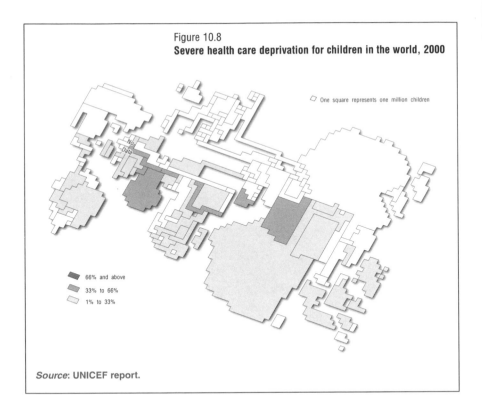

Figure 10.8
Severe health care deprivation for children in the world, 2000

One square represents one million children

No Data

66% and above
33% to 66%
1% to 33%

Source: UNICEF report.

'infections – intestinal', which is responsible for only 0.06% of deaths.) In Britain some (on average more affluent) people actually choose not to have their children immunised, as they have become less concerned about disease and more concerned about possible side-effects. Across the poor majority of the world some 15% of all children are severely deprived of health care, slightly more than of food. Only in Chad is a majority so deprived and in China less than 1% are so. Thus, as for education, the children of China do not feature on this map. In China the state has the power to ensure that almost all children receive very basic health care and education. Why this is not possible in the world's largest democracy – India – is a crucial question for those who believe that greater freedom brings greater prosperity to all. However, due partly to the widespread diffusion of immunisation as a means to prevent disease and to very basic health advice or treatment being available to 85% of the children in these countries, their numbers are growing rapidly.

Figure 10.9 presents one way in which the last seven measures can be combined to estimate where overall deprivation is most severe among the

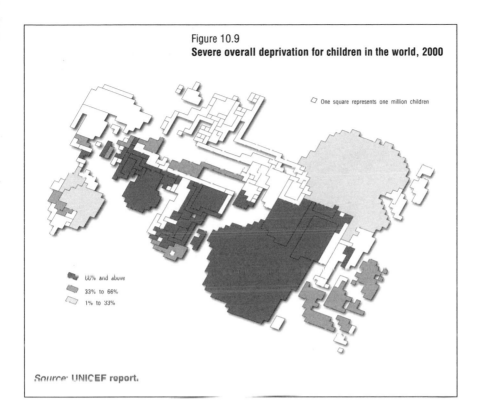

Figure 10.9
Severe overall deprivation for children in the world, 2000

One square represents one million children

60% and above
33% to 66%
1% to 33%

Source: UNICEF report.

45 countries shown here. Because the data used to measure deprivation came from surveys of households with children it is possible to calculate the proportion of children in each country who suffer from any one of the seven forms of severe deprivation as identified by the UNICEF report. By this measure over 90% of the children of the following countries suffer at least one such deprivation: Nepal, Ethiopia, Chad, Rwanda, Uganda, Burkina Faso, Benin, Bangladesh, Tanzania, Niger, Cambodia, Malawi and Mauritania. Only in the Philippines, South Africa, the Dominican Republic, Brazil, Colombia and China do more than half of all children escape such deprivation. Thus the majority of children in the poor majority of the world suffer from at least one form of severe deprivation. Rates are highest in Africa and the Indian subcontinent. Because the majority of the world's children live in these or similar countries and because all these measure err towards underestimating deprivation, it is safe to say that half the children in the world by 2000 lacked access to at least one of the following: decent water, sanitation, shelter, information, education, food and health care. Taken as a whole, as a global population, human beings have never had so much, never consumed so much, never

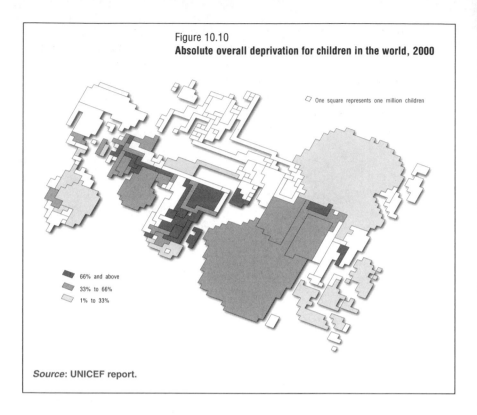

Figure 10.10
Absolute overall deprivation for children in the world, 2000

One square represents one million children

66% and above
33% to 66%
1% to 33%

Source: UNICEF report.

found treating water, building sewers, homes, providing news, schools, food or medicines so easy. The world is awash with the products of our collective labours. However, we have simultaneously managed to distribute these goods so that most children in the world do not have access to at least one of the most basic necessities of a decent life. Few in Britain would tolerate a single child in that country being deprived of any of these necessities and yet we tolerate the majority of children in the world growing up without them.

Where, though, do the poorest of the poor live? The causes of children living in severe deprivation are almost always due to a lack of resources or income. It is possible, however, that some children will suffer deprivation because of some form of discrimination, such as girls not being educated, or will suffer from stunted growth because they contracted a disease but did not live in the poorest of families. Thus if you are interested in an even more sure measure of deprivation, one that the authors of the UNICEF report termed 'absolute' deprivation, then with survey data you can measure the proportion

HUMAN GEOGRAPHY OF THE UK

of children suffering from at least two forms of severe deprivation. This measure also helps identify where the poorest or poor children live. Over a third, 37%, of children in these countries live in such absolute poverty. Over three-quarters of the children of Ethiopia, Nepal, Chad, Rwanda, Uganda, Niger, Burkina Faso, Tanzania and Mozambique live in such conditions. The poorest of the poor children of the world are mostly growing up in Africa. They live just a few thousand miles to the south of us, in a continent which the UK's 14 million children would fit into 33 times if you were to fit the cartogram of our island's children (see Figure 10.10) into that of the continent. For every child in the UK there are thus 33 children in Africa, most growing up in absolute poverty and almost all being severely deprived of at least one of the basic human needs. The surveys on which these figures were based involved interviewing the households of about one child in every 650 in Africa. This is a remarkable feat although surveying some the poorest countries, such as Somalia, was not possible. It is thus unlikely that such numbers overestimate the extent and depth of poverty on that continent.

An exercise

Mapping the human geography of the world is far more difficult than charting it within one rich nation for which there is abundant information. National censuses are often not taken in many countries, so the results of surveys can be hard to compare. Most importantly of all, international statistics are almost always presented for states which vary enormously in population and for which averages tend to disguise huge internal variations. The surveys which were used to generate the figures used to shade the maps in this chapter did differentiate almost universally between children living in rural or urban settlements and that alone showed that the worse conditions were concentrated out of the cities. Why else would so many people in poor countries be leaving the land for the cities?

One day soon maps will be drawn of the entire world's population that will differentiate both within countries as well as between them. Such maps will depict an even more starkly divided world than that shown in these pages on which, after all, the maps only depict up to 45 proportions. Suppose we had information on every square of the world cartogram, on the living conditions of every geographical concentration of 1 million children in the world. The maps would still be presenting averages of huge numbers of people, but for 2,150 regions rather than just 45 countries. How might such a map look? To begin draw an example for the 14 million children of the United Kingdom. First you need to draw your base map. You could use the outline drawn in the world maps in this book, in which the Kingdom is allocated 14 squares on the map in a

Figure 10.11
Example: regional cartogram of the children of the UK

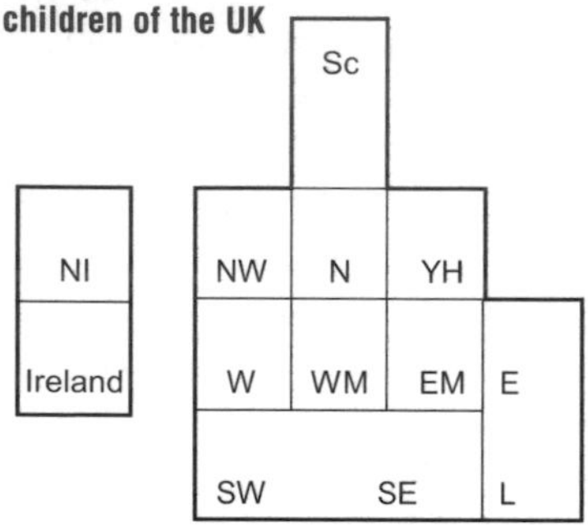

Children in 2001 (millions)

Northern Ireland	0.45	NI
North East	0.56	N
Wales	0.66	W
East Midlands	0.94	EM
South West	1.06	SW
Yorkshire and Humber	1.14	YH
East of England	1.21	E
West Midlands	1.23	WM
North West	1.57	NW
London	1.62	L
Scotland	1.65	Sc
South East	1.79	SE
UK	13.90	

Source: 2001 Census data.

shape which very roughly approximates its boundaries. A rough example is provided in Figure 10.11 where the regions and countries of the Kingdom have each been allocated a square, the South East and Scotland both with the largest numbers of children being allocated two. You need not use this example. Instead you could combine groups of six or seven of the constituencies used in this atlas. Their adults, as electorates, are listed at the end of Chapter 5, and one year's worth of their children are counted in Figure 1.3.

Having constructed your base map, each take one example from this book of a variable that can be calculated for your 14 areas. Remember that as each of the areas used in this book contain roughly equal populations, then in most cases you can simply sum proportions shown on these maps and divide them by the number of areas you are considering to calculate an average proportion which is representative of that group as a whole. Having drawn your map, compare it to the original map of variation over 84 or 85 areas. How much of the original detail do you lose? The answer to that question depends on how much variation within regions there was to begin with. A good example to take in the context of this chapter is employment in the financial industries, as depicted in Figure 8.2. To calculate the 2001 proportion in, say, London, simply sum all the proportions in the London squares (both the 1991 and the change proportions) and divide the result by 10, as there are ten squares in London. It should not take you too long to draw one map of what could be part of a future depiction of the human geography of the world workforce in financial industries.

Compare the maps that you have drawn and answer the following questions:

1 Which design appears to work best, is most visually appealing and presents the data both as simply and as accurately as appears possible?
2 Which design could most easily be extended to draw out a map of all the areas in the world in which 1 million children live?
3 How should such a world map be drawn, on where should it be centred, how should it be oriented (what should be at the top)?
4 To what extent does the loss of detail matter when this image is just part of a much larger picture? Does it really matter precisely which areas you combine to group a million children?
5 What information, if any, could you find to draw out variations in the globe in this way? Why is it so limited and will there be countries for the foreseeable future which will simply be labelled 'no data'.
6 What is the point of mapping the geography of people's lives in more detail? Does it serve a useful purpose or simply reinforce what we think we already know of our humanity and inhumanity?

CONCLUSION

In looking abroad we can begin to see the context within which the human geography of the United Kingdom is placed. There is very little which is tangible about the borders of this Kingdom and I have been liberal throughout these pages in variously referring to that entity, or Britain, or occasionally omitting Scotland, and at no point have I yet mentioned that the British are actually citizens of a European Union that now spans most of that continent, nor that their major industries and corporations are often as at home in New York as in London. A chapter on the world context to the human geography of this country could have discussed international forces, the now rapid movements of monies around the globe, international travel, the spread of the English language, the ever present fear of world recession or of future world wars of whatever kind. However, such an approach would have ignored the majority of the people of the world and in human geography it is still the human context which matters most. People are easily forgotten if you chart, map and describe the non-human worlds which influence our lives, the worlds of power, exploitation, profit and fear. All these worlds are created by people but to see the context within which we are placed simply in these terms reduces the majority of the world's people and especially the world's children to the sidelines. Fundamentally, the human geography of the world is made up of people and how they are distributed across the planet, how they will be so in the future, how they are living and what their millions of hopes, fears, needs and aspirations are, which is the context that most matters to the human geography of one small island. An island which in the past had a huge influence on most of the peoples of the rest of the world, which is reflected in the human geography of that island, and a past which almost certainly will matter more in the future.

Almost half the world was once governed directly or indirectly from ministries in London. Those colonies are now free but London's clerks, soldiers and civil servants have been replaced by an even greater army of financiers, bankers, insurers and underwriters. Along with New York (and to a far lesser extent Tokyo), London still exerts hugely undue influence on the world in proportion to its population. Furthermore, that influence is not planned, not thought out; it is the product of billions of financial transactions a year. Each transaction is a decision which alone appears rational, or at least a good guess, to move money or risk around in a way that is profitable for that moment in time. In aggregate and on average, all these transactions are extremely profitable for the people who make them, and there is increasingly little else underwriting the economy of the Kingdom. This increased profitability of finance is the key driving force that is altering the human geography of Britain; it is why the young and able are moving south in such numbers, and why lives to the north and west look more and more different. Within the UK it is the uneven distribution of the spoils of such financial gain which is the most important driver of growing social inequalities. And it matters where the profit is made. However, it matters also where it comes from. Money cannot be made out of thin air. Why do the bankers of the world come to London to raise their loans? Why does London house some of the insurers of last resort? What do we have which is so needed by the rest of the world that they pay us tithes, in effect, for our advice, time and acumen? After all, money is easily moved. Why have the world's bankers not moved to Mumbai, Rio or Hong Kong? London is often a cold, wet and dreary place to live, its empire long gone, its country stripped of its most easily extracted resources and most of its people are not living in a state of luxury. In the capital itself is found one of the largest concentrations of poverty in a rich nation in the world. Britain is no longer a military power of much substance, and most of its infrastructure, its buildings, sewers, railways and roads are old. Why here and why now?

A large part of the reason for the United Kingdom's continued success as a world banker is inertia. However, inertia cannot explain why that success is growing, why more and more of the world's money is flowing through London and why more and more people are employed there to help take a slice off it as it travels through. Of course most of the financial service workers of London do not work in international finance. They help cycle monies within the Kingdom in the form of pensions, insurance and loans, but these activities alone cannot generate extra money. It is the much smaller numbers of people who invest the pension funds, underwrite the insurance firms and raise the capital from outside the Kingdom to make millions of loans possible who actually bring money into the country. Increasingly, people with money in the rest of the world trust people in London to do this for them. They feel safer dealing with the largest of banks and companies, those which have offices in every major city and their headquarters in England or America. What then has happened over time to bolster this trust, to end up with companies and governments thousands of miles away increasingly doing business in a European capital which they know will extract a profit from them? Turn to look at the state of most of the children of the world as just one possible reason.

The majority of the children of the world live in a state of severe poverty, more than a third in absolute deprivation. In the countries where most of these children live, and in the regions around them, what does the future look like? How safe are the governments of these countries? How likely is war? How secure are their financial institutions? How safe are their peoples from disease, famine, crime and disorder? Then look again at the maps of world poverty among children in this chapter. In terms of people, where is the nation-state that is situated furthest from the poor majority who will be the adults of this century? Which country is furthest from areas coloured various shades of grey on these maps? Which country sits in the only ocean of the world, the North Atlantic, not to be bordered by coastlines where children lack water, sanitation, education, health care, food, information and shelter? Where would you put your money, do your deals, if you wanted both it and them to be safe in the future? London does not have a unique advantage in geographical isolation. There are many places around the world where finance is booming, often in very small islands less regulated by government control and where faster profits can be made. But London does benefit both from its past and its current position. Money comes to London because it flows away from where most people need it and where one day they might try to take it back.

11 Future

...concluding the Kingdom

The United Kingdom is made up of many small islands dominated by one still small island: Britain. Less than 1% of the world's population lives in the UK. Most of these people were born in just a few hundred maternity units and attended one or two of a few thousand secondary schools. Soon a majority will go to university, most of them to just a few dozen campuses. One-third live in the few largest conurbations. By the area names and boundaries used in this book (and given their areas on these pages in brackets), these conurbations are both quick to list and the first is almost as large as the rest: London (10), Greater Manchester (3), Birmingham (2), Merseyside (2), Bristol (1), Glasgow (1), Leeds (1), Leicester (1), Sheffield (1) and Tyne and Wear (1). There is great interest within some geography circles in the Kingdom as to which city and conurbation holds what rank in any urban order – where their precise boundaries end and so on. However, a crude a list as that just given is as good as any. What matters is that this is all a small place. The domination by one conurbation is far more important than the scrabbling around of others to vie to be recognised as being in second, third and tenth place.

Most people in the United Kingdom live within 100 miles of half the population of the main island. This has been the case for many decades and, although the human geography of the UK is slowly changing, what characterises it as most unusual internationally is how slow that change has been for most of the last 100 years. Many places are still populated because they were populated in the distant past and homes were built there. Other land is relatively sparsely populated. In the south of England this is usually because it was sparsely populated in the past and that past is being preserved (rather than a lack of demand to live there). To the north and west large amounts of land are designated National Parks specifically to fossilise the visual appearance of their human-created landscapes. And it is not just here that time appears to be passing slowly. Outside the countryside most cities, towns, suburbs and even streets which were rich or poor a century ago are in much the same relative position now. Relatively stable population growth and then stagnation, along with strict planning controls and, most importantly, a remarkable degree of snobbishness (among those who can choose where to live), and a sense that place and location matters, have helped to ensure that it does, and does so more now than place has mattered for many decades.

The majority of variation across the human landscape of the UK can be revealed by dividing that landscape up into just the seven dozen areas used in this book and treating Northern Ireland as an additional slightly bigger area, as those who designed the European constituencies chose to do. The Province has mainly been described with the words 'no data' in these pages and, although unfortunate from the point of view of people interested in its human geography, that perhaps best still describes it in terms of both how it is viewed from Britain and practically in how the British state tends to exclude it from its national datasets. It is a bit cheeky to claim that this book is a human geography of the UK given how little data on the Province is included unless you sign up to the argument that this is how the Province is mostly represented within the UK – as not being there. Turning to the mainland, the remaining 84 areas for which we mostly do have comparable data are a relatively arbitrary subdivision of Britain. The subdivision is unique as a regional description of the country only in that, unlike any other, it divides the population up into equally sized groups. That has great advantages in revealing the main contours of the landscape. Similarly, the presentation of variation of these areas as a visual landscape is also novel.

Most depictions of the human geography of the UK concentrate on just 12 or so regions and countries. The standard regions (or Government Office Regions) are areas which deliberately mix up town and country and rich and poor, and thus hide the majority of the variation in the landscape. Over the last couple of decades a dozen atlases of the human geography of parts of the UK have been drawn at finer spatial scales, but most of the variation they show is also revealed in these pages. It is between these 85 areas that the lines that are most crucial to dividing up the population of the UK roughly lie. The patterns to life in the UK do become more detailed, more fascinating and increasingly more revealing as smaller and smaller groups of people are mapped, but such pictures are merely elaborations on the basic spatial social divisions revealed here. Within each European constituency another 85 or so areas could be drawn, showing, in most, variation akin to that found nationally but with divides that were less stark. That exercise could be repeated again, resulting in some 600,000 areas each containing roughly 85 adults.

Only at such a fine level of detail would we begin to see a different nature to the human landscape emerge. Individual enclosures of extreme affluence and poverty would become clear, but most of the country would appear more uniform. If you take two short streets in the UK at random, each containing about 85 people, they are unlikely to differ that much in their social characteristics. It is not until you group people in very large numbers by where they live that for most you begin to see the importance of place everywhere.

Accounts of the human geography of the UK, just like this one, tend to concentrate on what most differentiates people, the peaks and troughs of the landscape, its steepest slopes, cliffs and other disjunctures. However, a great deal of the landscape is, in detail, quite uniform even if, over time, becoming a little less so. Even on the simple maps in this book your eye is naturally drawn to try to see what is most interesting and to ignore the ways in which most places are nearly average. However, this underlying uniformity, the slight social differences revealed within it and the sharp divides that consistently separate a few places from the rest all require the same careful and continuous maintenance for this landscape to be preserved as it is. People in the UK, over a long period of time have to behave and be controlled in very consistent ways to keep the suburbs nondescript, the inner cities mostly poor, the home counties most affluent and the periphery peripheral. Historically children must act much like their parent(s) did before them, socially people must behave like their peers, economically cohorts must act lemming-like in reacting to what collectively hits them due to the timing of their births, and politically people must act a little more like their neighbours than they otherwise might if our maps of the landscape are to continue to look neat.

Were future generations to behave very differently from their elders as they age, were people's actions not largely predictable from those of their social peers, were cohorts not to accept their historical fate and neighbours their spatial obligations to act similarly to one another, it would become harder and harder to describe the landscape. To extend the physical metaphor, it would be like water beginning to flow uphill, mountains to come crashing down, smooth slopes to become pitted and unchartable. Fortunately, if only for the cartographers of this particular landscape, the British population do not go in for revolution, their governments occasionally try more enthusiastically than usual to prop up areas which are faring poorly and curtail the extravagancies in places where affluence amasses but in the main we (in our actions) and they (in their Acts) function as gardeners maintaining the look of a landscape largely created in Victorian times. We produce roughly the right numbers of babies to fill the homes of the future (at lower densities), largely bring our children up where and how we are supposed to, fill almost all of the school and university places allocated to us, take the entry-level jobs on offer, buy homes when we can as soon as we can afford to, retire or drop out as ill when we are told or feel we are no longer needed, and finally move compliantly in our dotages away from places where the young are most needed (and for those of us with money leave much of it to our families so they can afford to buy where we once lived). Were any of these patterns to change, much of the landscape would quickly erode and reform.

Internally to the UK there are a few trends which could be highlighted for those interested in how the future geography of the UK might differ. Fewer children are being born per woman than at any time in the last century and later in most women's lives. More women than men are now receiving the 'best' qualifications and many more young men have left the country than we realised before we took the latest Census. Many more children are being educated to higher levels than ever before, although still in such a way that they are socially sorted much as their parents were. Today's children and now young adults are simply tested far more, given more certificates and made to wait a few more years before being allowed entry into the workforce. There is change, it is true, in that they are entering a workforce now more divided geographically, socially and economically than that in which their parent(s) laboured. More jobs are now either menial or labelled as professional than in the past and remunerated more unequally accordingly. This enhanced division of labour is being reflected in the landscape as the price of entry into many places, through house prices, diverges. It is reflected in new layers in the sediment beneath the surface as patterns to our mortality become ever more geographically distinct. As more people die than are being born, higher numbers are being allowed to emigrate to the UK from abroad, and so another future change that could be highlighted is in people's more disparate geographical origins. But the UK has, when it has suited it, allowed and encouraged similar proportions in from aboard before.

If anything, it is external influences on the UK which are most likely to alter the landscape in the future and which, it could be argued, did most to determine its shape in the past and present. The south is richer basically because of where it is and hence where it was and is nearer to. Place the UK in the Mediterranean sea and it would be near the mountainous north where the capital would lie. The ten cities listed at the start of this conclusion grew up on the profits of organising empire, and on cotton, shipping, wool, steel, coal and iron – their export, import and commodification. The two great depressions of the last century, separated by 50 years, were worldwide events, the impacts of which are still etched in the human landscape of the UK. Similarly, the legacies of two world wars are still felt in numerous ways, most obviously as the survivors of the million babies born in celebration in 1946 turn 60 in 2006. What happens to commuter-belt suburbs when such a large group no longer needs to live there will be interesting, especially as the young are choosing or being forced to live for longer in city centres – another global trend. The demand for the labour and skills of the young in these cities is increasingly financed internationally too. We consume far more than we physically produce and can only do so because of the financial services we sell abroad. Were international trust in the UK – and in London in particular – to fall, the result would, in the short term at least, be devastating. Only just over half a percentage of the world's children live in the UK as compared to its percentage point of the world's adults and thus the Kingdom is set to halve in size on the global map of our human landscape. Much more is both going on and changing outside as compared to within.

Although the country is a tiny speck on the global landscape, and is set to become smaller still, it holds a position of disproportionate importance on that world map. This is not just through its impact in consuming far more than it produces, in its influence on far-away nations through world markets, through its continued military adventures or its luck in being home to the language of international trade. The UK is also disproportionately important in being the setting for one of the longest-running social experiments in the world. What happens when you allow people to elect their masters over a long period of time, when your policies to shape the landscape have not been interrupted by invasion or revolution, when your population has been largely stable in location, when your wealth has been great, when you have lived through generation after generation of social reform, when you have paid pensions for almost a century, provided free health care for over half a century, when you have had most of the advantages of location (worldwide) for longer than almost anywhere else and have had an empire to help pay for your projects until recently? What occurs in these almost the best of circumstances is an experiment in tempering capitalism. Our landscape reveals how a population organised through markets and moderated by its government behaves. This is how people become sorted out over space if you organise life in such a way. What we can see now and should expect in the future is perhaps the best that we can expect to get – if this is how we collectively choose to behave. It's not bad, it's not great, it's not fair, it's no utopia, but for most it is a pleasant enough brick, concrete and asphalt land; it is planned, maintained, in places agonised over and preserved. It's a very British landscape, a southern English garden with rough edges, sitting in a far less clearly ordered world, the disorder of which it is disproportionately responsible for, and which will matter most for its future in the years to come.

Appendix:
the places mapped in this book

This appendix defines each European constituency used in this book in terms of the (approximate) local authority areas which it includes and the (exact) Westminster constituencies included as their boundaries existed around 2001.

Note: LA = local authority, LB = London borough, UA = unitary authority, MB = metropolitan borough, CA = council area

No.	European constituency	Local authorities	Westminster constituencies
		London	
1	London Central	Camden LB; City of London LB; Hammersmith and Fulham LB; Islington LB; Kensington and Chelsea LB; Westminster LB.	Cities of London and Westminster; Hammersmith and Fulham; Hampstead and Highgate; Holborn and St Pancras; Islington North; Islington South and Finsbury; Kensington and Chelsea; Regent's Park and Kensington North.
2	London East	Barking and Dagenham LB; Havering LB; Redbridge LB.	Barking; Dagenham; East Ham; Hornchurch; Ilford North; Ilford South; Romford; Upminster.
3	London North	Barnet LB; Enfield LB; Haringey LB.	Chipping Barnet; Edmonton; Enfield North; Enfield, Southgate; Finchley and Golders Green; Hendon; Hornsey and Wood Green; Tottenham.

(Continued)

(Continued)

No.	European constituency	Local authorities	Westminster constituencies
4	**London North East**	Hackney LB; Newham LB; Tower Hamlets LB; Waltham Forest LB.	Bethnal Green and Bow; Chingford and Woodford Green; Hackney North and Stoke Newington; Hackney South and Shoreditch; Leyton and Wanstead; Poplar and Canning Town; Walthamstow; West Ham.
5	**London North West**	Brent LB; Harrow LB; Hillingdon LB.	Brent East; Brent North; Brent South; Harrow East; Harrow West; Hayes and Harlington; Ruislip–Northwood; Uxbridge.
6	**London South & Surrey East**	Croydon LB; Epsom and Ewell LA; Sutton LB; Tandridge LA.	Carshalton and Wallington; Croydon Central; Croydon North; Croydon South; East Surrey; Epsom and Ewell; Sutton and Cheam.
7	**London South East**	Bexley LB; Bromley LB; Greenwich LB.	Beckenham; Bexleyheath and Crayford; Bromley and Chislehurst; Eltham; Erith and Thamesmead; Greenwich and Woolwich; Old Bexley and Sidcup; Orpington.
8	**London South Inner**	Lambeth LB; Lewisham LB; Southwark LB.	Camberwell and Peckham; Dulwich and West Norwood; Lewisham East; Lewisham West; Lewisham, Deptford; North Southwark and Bermondsey; Streatham; Vauxhall.
9	**London South West**	Kingston upon Thames LB; Merton LB; Wandsworth LB.	Battersea; Kingston and Surbiton; Mitcham and Morden; Putney; Richmond Park; Tooting; Wimbledon.
10	**London West**	Ealing LB; Hounslow LB; Richmond upon Thames LB; Spelthorne LA.	Brentford and Isleworth; Ealing North; Ealing, Acton and Shepherd's Bush; Ealing, Southall; Feltham and Heston; Spelthorne; Twickenham.

(Continued)

APPENDIX: THE PLACES MAPPED IN THIS BOOK

(Continued)

No.	European constituency	Local authorities	Westminster constituencies
		South East	
11	**Buckinghamshire & Oxfordshire East**	Aylesbury Vale LA; Cherwell LA; Chiltern LA; South Bucks LA; South Oxfordshire LA; Wycombe LA.	Aylesbury; Banbury; Beaconsfield; Buckingham; Chesham and Amersham; Henley; Wycombe.
12	**East Sussex & Kent South**	Brighton and Hove UA; Eastbourne LA; Hastings LA; Lewes LA; Rother LA; Tunbridge Wells LA; Wealden LA.	Bexhill and Battle; Brighton, Kemptown; Brighton, Pavilion; Eastbourne; Hastings and Rye; Lewes; Tunbridge Wells; Wealden.
13	**Hampshire North & Oxford**	Basingstoke and Deane LA; Oxford LA; Vale of White Horse LA; West Berkshire UA; West Oxfordshire LA.	Basingstoke; Newbury; North West Hampshire; Oxford East; Oxford West and Abingdon; Wantage; Witney.
14	**Kent East**	Ashford LA; Canterbury LA; Dover LA; Shepway LA; Swale LA; Thanet LA.	Ashford; Canterbury; Dover; Faversham and Mid Kent; Folkestone and Hythe; North Thanet; Sittingbourne and Sheppey; South Thanet.
15	**Kent West**	Dartford LA; Gravesham LA; Maidstone LA; Medway UA; Sevenoaks LA; Tonbridge and Malling LA.	Chatham and Aylesford; Dartford; Gillingham; Gravesham; Maidstone and The Weald; Medway; Sevenoaks; Tonbridge and Malling.
16	**South Downs West**	Arun LA; Chichester LA; East Hampshire LA; Hart LA; Rushmoor LA; Waverley LA; Winchester LA.	Aldershot; Bognor Regis and Littlehampton; Chichester; East Hampshire; North East Hampshire; South West Surrey; Winchester.

(Continued)

APPENDIX: THE PLACES MAPPED IN THIS BOOK

(Continued)

No.	European constituency	Local authorities	Westminster constituencies
17	**Surrey**	Elmbridge LA; Guildford LA; Mole Valley LA; Reigate and Barnstead LA; Runnymede LA; Surrey Heath LA; Woking LA.	Esher and Walton; Guildford; Mole Valley; Reigate; Runnymede and Weybridge; Surrey Heath; Woking.
18	**Sussex West**	Adur LA; Crawley LA; Horsham LA; Mid Sussex LA; Worthing LA.	Arundel and South Downs; Crawley; East Worthing and Shoreham; Horsham; Hove; Mid Sussex; Worthing West.
19	**Thames Valley**	Bracknell Forest UA; Reading UA; Slough UA; Windsor and Maidenhead UA; Wokingham UA.	Bracknell; Maidenhead; Reading East; Reading West; Slough; Windsor; Wokingham.
20	**Wight & Hampshire South**	Eastleigh LA; Fareham LA; Gosport LA; Havant LA; Isle of Wight UA; Portsmouth UA.	Eastleigh; Fareham; Gosport; Havant; Isle of Wight; Portsmouth North; Portsmouth South.
		South West	
21	**Bristol**	Bristol, City of UA; South Gloucestershire UA.	Bristol East; Bristol North West; Bristol South; Bristol West; Kingswood; Northavon; Woodspring.
22	**Cornwall & West Plymouth**	Caradon LA; Carrick LA; Isles of Scilly LA; Kerrier LA; North Cornwall LA; Penwith LA; Plymouth UA; Restormel LA.	Falmouth and Camborne; North Cornwall; Plymouth, Devonport; Plymouth, Sutton; South East Cornwall; St Ives; Truro and St Austell.

(Continued)

(Continued)

No.	European constituency	Local authorities	Westminster constituencies
23	**Devon & East Plymouth**	Exeter LA; Mid Devon LA; South Hams LA; Teignbridge LA; Torbay UA; Torridge LA; West Devon LA.	Exeter; South West Devon; Teignbridge; Tiverton and Honiton; Torbay; Torridge and West Devon; Totnes.
24	**Dorset & East Devon**	Bournemouth UA; East Devon LA; East Dorset LA; North Dorset LA; Poole UA; Purbeck LA; West Dorset LA; Weymouth and Portland LA.	Bournemouth East; Bournemouth West; East Devon; Mid Dorset and North Poole; North Dorset; Poole; South Dorset; West Dorset.
25	**Gloucestershire**	Cheltenham LA; Cotswold LA; Forest of Dean LA; Gloucester LA; Malvern Hills LA; Stroud LA; Tewkesbury LA.	Cheltenham; Cotswold; Forest of Dean; Gloucester; Stroud; Tewkesbury; West Worcestershire.
26	**Itchen, Test & Avon**	Christchurch LA; New Forest LA; Salisbury LA; Southampton UA; Test Valley LA.	Christchurch; New Forest East; New Forest West; Romsey; Salisbury; Southampton, Itchen; Southampton, Test.
27	**Somerset & North Devon**	Mendip LA; North Devon LA; North Somerset UA; Sedgemoor LA; South Somerset LA; Taunton Deane LA; West Somerset LA.	Bridgwater; North Devon; Somerton and Frome; Taunton; Wells; Weston-Super-Mare; Yeovil.
28	**Wiltshire North & Bath**	Bath and North East Somerset UA; Kennet LA; North Wiltshire LA; Swindon UA; West Wiltshire LA.	Bath; Devizes; North Swindon; North Wiltshire; South Swindon; Wansdyke; Westbury.

(Continued)

(Continued)

No.	European constituency	Local authorities	Westminster constituencies
		East of England	
29	**Bedfordshire & Milton Keynes**	Bedford LA; Luton UA; Mid Bedfordshire LA; Milton Keynes UA; South Bedfordshire LA.	Bedford; Luton North; Luton South; Mid Bedfordshire; Milton Keynes South West; North East Bedfordshire; North East Milton Keynes; South West Bedfordshire.
30	**Cambridgeshire**	Cambridge LA; East Cambridgeshire LA; Fenland LA; Huntingdonshire LA; Peterborough UA; South Cambridgeshire LA.	Cambridge; Huntingdon; North East Cambridgeshire; North West Cambridgeshire; Peterborough; South Cambridgeshire; South East Cambridgeshire.
31	**Essex North & Suffolk South**	Babergh LA; Braintree LA; Chelmsford LA; Colchester LA; Maldon LA; Tendring LA.	Braintree; Colchester; Harwich; Maldon and East Chelmsford; North Essex; South Suffolk; West Chelmsford.
32	**Essex South**	Basildon LA; Castle Point LA; Rochford LA; Southend-on-Sea UA; Thurrock UA.	Basildon; Billericay; Castle Point; Rayleigh; Rochford and Southend East; Southend West; Thurrock.
33	**Essex West & Hertfordshire East**	Brentwood LA; Broxbourne LA; East Hertfordshire LA; Epping Forest LA; Harlow LA; North Hertfordshire LA; Stevenage LA; Uttlesford LA.	Brentwood and Ongar; Broxbourne; Epping Forest; Harlow; Hertford and Stortford; North East Hertfordshire; Saffron Walden; Stevenage.
34	**Hertfordshire**	Dacorum LA; Hertsmere LA; St Albans LA; Three Rivers LA; Watford LA; Welwyn Hatfield LA.	Hemel Hempstead; Hertsmere; Hitchin and Harpenden; South West Hertfordshire; St Albans; Watford; Welwyn Hatfield.

(Continued)

APPENDIX: THE PLACES MAPPED IN THIS BOOK

(Continued)

No.	European constituency	Local authorities	Westminster constituencies
35	Norfolk	Broadland LA; Great Yarmouth LA; King's Lynn and West Norfolk LA; North Norfolk LA; Norwich LA; South Norfolk LA.	Great Yarmouth; Mid Norfolk; North Norfolk; North West Norfolk; Norwich North; Norwich South; South Norfolk.
36	Suffolk & South West Norfolk	Breckland LA; Forest Heath LA; Ipswich LA; Mid Suffolk LA; St Edmundsbury LA; Suffolk Coastal LA; Waveney LA.	Bury St Edmunds; Central Suffolk and North Ipswich; Ipswich; South West Norfolk; Suffolk Coastal; Waveney; West Suffolk.
	West Midlands		
37	Birmingham East	Birmingham MB.	Birmingham, Edgbaston; Birmingham, Hall Green; Birmingham, Hodge Hill; Birmingham, Ladywood; Birmingham, Northfield; Birmingham, Selly Oak; Birmingham, Sparkbrook and Small Heath; Birmingham, Yardley.
38	Birmingham West	Sandwell MB; Walsall MB.	Aldridge–Brownhills; Birmingham, Erdington; Birmingham, Perry Barr; Sutton Coldfield; Walsall North; Walsall South; West Bromwich East; West Bromwich West.
39	Coventry & North Warwickshire	Coventry MB; North Warwickshire LA; Nuneaton and Bedworth LA; Solihull MB.	Coventry North East; Coventry North West; Coventry South; Meriden; North Warwickshire; Nuneaton; Solihull.

(Continued)

(Continued)

No.	European constituency	Local authorities	Westminster constituencies
40	Herefordshire & Shropshire	Bridgnorth LA; Herefordshire, County of UA; North Shropshire LA; Oswestry LA; Shrewsbury and Atcham LA; South Shropshire LA; Telford and Wrekin UA; Wyre Forest LA.	Hereford; Leominster; Ludlow; North Shropshire; Shrewsbury and Atcham; Telford; The Wrekin; Wyre Forest.
41	Midlands West	Dudley MB; Wolverhampton MB.	Dudley North; Dudley South; Halesowen and Rowley Regis; Stourbridge; Warley; Wolverhampton North East; Wolverhampton South East; Wolverhampton South West.
42	Staffordshire East & Derby	Cannock Chase LA; Derby UA; East Staffordshire LA; Lichfield LA; South Derbyshire LA; Tamworth LA.	Burton; Cannock Chase; Derby North; Derby South; Lichfield; South Derbyshire; Tamworth.
43	Staffordshire West & Congleton	Congleton LA; Newcasle-under-Lyme LA; South Staffordshire LA; Stafford LA; Stoke-on-Trent UA.	Congleton; Newcastle-under-Lyme; South Staffordshire; Stafford; Stoke-on-Trent Central; Stoke-on-Trent North; Stoke-on-Trent South; Stone.
44	Worcestershire & South Warwickshire	Bromsgrove LA; Redditch LA; Rugby LA; Stratford-upon-Avon LA; Warwick LA; Worcester LA; Wychavon LA.	Bromsgrove; Mid Worcestershire; Redditch; Rugby and Kenilworth; Stratford-on-Avon; Warwick and Leamington; Worcester.

(Continued)

(Continued)

No.	European constituency	Local authorities	Westminster constituencies
		East Midlands	
45	**Leicester**	Harborough LA; Leicester UA; Melton LA; Oadby and Wigston LA; Rutland UA; South Kesteven LA.	Charnwood; Grantham and Stamford; Harborough; Leicester East; Leicester South; Leicester West; Rutland and Melton.
46	**Lincolnshire**	Boston LA; East Lindsey LA; Lincoln LA; North East Lincolnshire UA; North Kesteven LA; South Holland LA; West Lindsey LA.	Boston and Skegness; Cleethorpes; Gainsborough; Great Grimsby; Lincoln; Louth and Horncastle; Sleaford and North Hykeham; South Holland and the Deepings.
47	**Northampton-shire & Blaby**	Blaby LA; Corby LA; Daventry LA; East Northamptonshire LA; Kettering LA; Northampton LA; South Northamptonshire LA; Wellingborough LA.	Blaby; Corby; Daventry; Kettering; Northampton North; Northampton South; Wellingborough.
48	**Nottingham & Leicestershire North West**	Charnwood LA; Gedling LA; Hinckley and Bosworth LA; North West Leicestershire LA; Nottingham UA; Rushcliffe LA.	Bosworth; Gedling; Loughborough; North West Leicestershire; Nottingham East; Nottingham North; Nottingham South; Rushcliffe.
49	**Nottingham-shire North & Chesterfield**	Bassetlaw LA; Bolsover LA; Chesterfield LA; Mansfield LA; Newark and Sherwood LA; North East Derbyshire LA.	Bassetlaw; Bolsover; Chesterfield; Mansfield; Newark; North East Derbyshire; Sherwood.

(Continued)

(Continued)

No.	European constituency	Local authorities	Westminster constituencies
50	**Peak District**	Amber Valley LA; Ashfield LA; Broxtowe LA; Derbyshire Dales LA; Erewash LA; High Peak LA; Staffordshire Moorlands LA.	Amber Valley; Ashfield; Broxtowe; Erewash; High Peak; Staffordshire Moorlands; West Derbyshire.
		North West	
51	**Cheshire East**	Halton UA; Macclesfield LA; Vale Royal LA; Warrington UA.	Altrincham and Sale West; Halton; Macclesfield; Tatton; Warrington North; Warrington South; Weaver Vale.
52	**Cheshire West & Wirral**	Chester LA; Crewe and Nantwich LA; Ellesmere Port and Neston LA; Wirral MB.	Birkenhead; City of Chester; Crewe and Nantwich; Eddisbury; Ellesmere Port and Neston; Wallasey; Wirral South; Wirral West.
53	**Cumbria & Lancashire North**	Allerdale LA; Barrow-in-Furness LA; Carlisle LA; Copeland LA; Eden LA; Lancaster LA; South Lakeland LA; Wyre LA.	Barrow and Furness; Carlisle; Copeland; Lancaster and Wyre; Morecambe and Lunesdale; Penrith and The Border; Westmorland and Lonsdale; Workington.
54	**Greater Manchester Central**	Manchester MB; Stockport MB.	Cheadle; Hazel Grove; Manchester Central; Manchester, Blackley; Manchester, Gorton; Manchester, Withington; Stockport; Wythenshawe and Sale East.
55	**Greater Manchester East**	Oldham MB; Rochdale MB; Tameside MB.	Ashton under Lyne; Denton and Reddish; Heywood and Middleton; Oldham East and Saddleworth; Oldham West and Royton; Rochdale; Stalybridge and Hyde.

(Continued)

APPENDIX: THE PLACES MAPPED IN THIS BOOK

(Continued)

No.	European constituency	Local authorities	Westminster constituencies
56	Greater Manchester West	Bolton MB; Salford MB; Trafford MB.	Bolton North East; Bolton South East; Bolton West; Bury South; Eccles; Salford; Stretford and Urmston; Worsley.
57	Lancashire Central	Blackpool UA; Burnley LA; Fylde LA; Pendle LA; Preston LA; Ribble Valley LA.	Blackpool North and Fleetwood; Blackpool South; Burnley; Fylde; Pendle; Preston; Ribble Valley
58	Lancashire South	Blackburn with Darwen UA; Bury MB; Chorley LA; Hyndburn LA; Rossendale LA; South Ribble LA; West Lancashire LA.	Blackburn; Bury North; Chorley; Hyndburn; Rossendale and Darwen; South Ribble; West Lancashire.
59	Merseyside East & Wigan	Knowsley MB; St Helens MB; Wigan MB.	Knowsley North and Sefton East; Knowsley South; Leigh; Makerfield; St Helens North; St Helens South; Wigan.
60	Merseyside West	Liverpool MB; Sefton MB.	Bootle; Crosby; Liverpool, Garston; Liverpool, Riverside; Liverpool, Walton; Liverpool, Wavertree; Liverpool, West Derby; Southport.
		Yorkshire and the Humber	
61	East Yorkshire & North Lincolnshire	East Riding of Yorkshire UA; Kingston upon Hull, City of UA; North Lincolnshire UA.	Beverley and Holderness; Brigg and Goole; East Yorkshire; Haltemprice and Howden; Kingston upon Hull East; Kingston upon Hull North; Kingston upon Hull West and Hessle; Scunthorpe.
62	Leeds	Leeds MB.	Elmet; Leeds Central; Leeds East; Leeds North East; Leeds North West; Leeds West; Morley and Rothwell; Pudsey.

(Continued)

(Continued)

No.	European constituency	Local authorities	Westminster constituencies
63	**North Yorkshire**	Craven LA; Harrogate LA; Ryedale LA; Scarborough LA; Selby LA; York UA.	City of York; Harrogate and Knaresborough; Ryedale; Scarborough and Whitby; Selby; Skipton and Ripon; Vale of York.
64	**Sheffield**	Sheffield MB.	Barnsley West and Penistone; Sheffield Central; Sheffield, Attercliffe; Sheffield, Brightside; Sheffield, Hallam; Sheffield, Heeley; Sheffield, Hillsborough.
65	**Yorkshire South**	Barnsley MB; Doncaster MB; Rotherham MB.	Barnsley Central; Barnsley East and Mexborough; Don Valley; Doncaster Central; Doncaster North; Rother Valley; Rotherham; Wentworth.
66	**Yorkshire South West**	Kirklees MB; Wakefield MB.	Batley and Spen; Colne Valley; Dewsbury; Hemsworth; Huddersfield; Normanton; Pontefract and Castleford; Wakefield.
67	**Yorkshire West**	Bradford MB; Calderdale MB.	Bradford North; Bradford South; Bradford West; Calder Valley; Halifax; Keighley; Shipley.
		North East	
68	**Cleveland & Richmond**	Hambleton LA; Hartlepool UA; Middlesbrough UA; Redcar and Cleveland UA; Richmondshire LA; Stockton-on-Tees UA.	Hartlepool; Middlesbrough; Middlesbrough South and East Cleveland; Redcar; Richmond (Yorks.); Stockton North; Stockton South.
69	**Durham**	Chester-le-Street LA; Darlington UA; Derwentside LA; Durham LA; Easington LA; Sedgefield LA; Teesdale LA; Wear Valley LA.	Bishop Auckland; Blaydon; City of Durham; Darlington; Easington; North Durham; North West Durham; Sedgefield.

(Continued)

APPENDIX: THE PLACES MAPPED IN THIS BOOK

(Continued)

No.	European constituency	Local authorities	Westminster constituencies
70	**Northumbria**	Alnwick LA; Berwick-upon-Tweed LA; Blyth Valley LA; Castle Morpeth LA; Newcastle upon Tyne MB; North Tyneside MB; Tynedale LA; Wansbeck LA.	Berwick-upon-Tweed; Blyth Valley; Hexham; Newcastle upon Tyne Central; Newcastle upon Tyne North; North Tyneside; Tynemouth; Wansbeck.
71	**Tyne & Wear**	Gateshead MB; South Tyneside MB; Sunderland MB.	Gateshead East and Washington West; Houghton and Washington East; Jarrow; Newcastle upon Tyne East and Wallsend; South Shields; Sunderland North; Sunderland South; Tyne Bridge.
		Wales	
72	**Mid & West Wales**	Carmarthenshire UA; Ceredigion UA; Pembrokeshire UA; Powys UA.	Brecon and Radnorshire; Carmarthen East and Dinefwr; Carmarthen West and South Pembrokeshire; Ceredigion; Llanelli; Meirionnydd Nant Conwy; Montgomeryshire; Preseli Pembrokeshire.
73	**North Wales**	Conwy UA; Denbighshire UA; Flintshire UA; Gwynedd UA; Isle of Anglesey UA; Wrexham UA.	Alyn and Deeside; Caernarfon; Clwyd South; Clwyd West; Conwy; Delyn; Vale of Clwyd; Wrexham; Ynys Mon.
74	**South Wales Central**	Cardiff UA; Rhondda, Cynon, Taff UA; Vale of Glamorgan UA.	Cardiff Central; Cardiff North; Cardiff South and Penarth; Cardiff West; Cynon Valley; Pontypridd; Rhondda; Vale of Glamorgan.
75	**South Wales East**	Blaenau Gwent UA; Caerphilly UA; Merthyr Tydfil UA; Monmouthshire UA; Newport UA; Torfaen UA.	Blaenau Gwent; Caerphilly; Islwyn; Merthyr Tydfil and Rhymney; Monmouth; Newport East; Newport West; Torfaen.

(Continued)

(Continued)

No.	European constituency	Local authorities	Westminster constituencies
76	South Wales West	Bridgend UA; Neath Port Talbot UA; Swansea UA.	Aberavon; Bridgend; Gower; Neath; Ogmore; Swansea East; Swansea West.

<table>
<tr><td colspan="4">Scotland</td></tr>
</table>

No.	European constituency	Local authorities	Westminster constituencies
77	Central Scotland	East Ayrshire CA; Falkirk CA; North Lanarkshire CA; South Lanarkshire CA.	Airdrie and Shotts; Coatbridge and Chryston; Cumbernauld and Kilsyth; East Kilbride; Falkirk East; Falkirk West; Hamilton North and Bellshill; Hamilton South; Kilmarnock and Loudoun; Motherwell and Wishaw.
78	Glasgow	Glasgow City CA.	Glasgow Anniesland; Glasgow Baillieston; Glasgow Cathcart; Glasgow Govan; Glasgow Kelvin; Glasgow Maryhill; Glasgow Pollok; Glasgow Rutherglen; Glasgow Shettleston; Glasgow Springburn.
79	Highlands & Islands	Argyll and Bute CA; Eilean Siar CA; Highland CA; Moray CA; Orkney Islands CA; Shetland Islands CA.	Argyll and Bute; Caithness, Sutherland and Easter Ross; Inverness East, Nairn and Lochaber; Moray; Orkney and Shetland; Ross, Skye and Inverness West; Western Isles.
80	Lothian	Edinburgh, City of CA; Midlothian CA; West Lothian CA.	Edinburgh Central; Edinburgh East and Musselburgh; Edinburgh North and Leith; Edinburgh Pentlands; Edinburgh South; Edinburgh West; Linlithgow; Livingston; Midlothian.
81	Mid Scotland & Fife	Clackmannanshire CA; Fife CA; Perth and Kinross CA; Stirling CA.	Central Fife; Dunfermline East; Dunfermline West; Kirkaldy; North East Fife; North Tayside; Ochil; Perth; Stirling.

(Continued)

APPENDIX: THE PLACES MAPPED IN THIS BOOK

No.	European constituency	Local authorities	Westminster constituencies
82	North East Scotland	Aberdeen City CA; Aberdeenshire CA; Angus CA; Dundee City CA.	Aberdeen Central; Aberdeen North; Aberdeen South; Angus; Banff and Buchan; Dundee East; Dundee West; Gordon; West Aberdeenshire and Kincardine.
83	South of Scotland	Dumfries and Galloway CA; East Lothian CA; North Ayrshire CA; Scottish Borders CA; South Ayrshire CA.	Ayr; Carrick, Cumnock and Doon Valley; Clydesdale; Cunninghame South; Dumfries; East Lothian; Galloway and Upper Nithsdale; Roxburgh and Berwickshire; Tweedale, Ettrick and Lauderdale.
84	West of Scotland	East Dunbartonshire CA; East Renfrewshire CA; Inverclyde CA; Renfrewshire CA; West Dunbartonshire CA.	Clydebank and Milngavie; Cunninghame North; Dumbarton; Eastwood; Greenock and Inverclyde; Paisley North; Paisley South; Strathkelvin and Bearsden; West Renfrewshire.
		Northern Ireland	
85	Northern Ireland	All in Northern Ireland	All in Northern Ireland

Brief reference list

This book was written for students at university studying at the start of the twenty-first century, assumed to be living in, or interested in the United Kingdom. Given this audience it is assumed that they are largely computer literate and could search for sources or further information on the world wide web. It was also assumed that they would not welcome the insertion in this text of numerous references to printed works, many of which could only be found in a few university libraries. The main sources that were used in writing this book and drawing the maps shown here were:

The population censuses of 1981, 1991 and 2001 (http://www.mimas.ac.uk/lct/).

Social Trends (http://www.statistics.gov.uk/.

The House of Commons Library web pages (in particular for the definitions of the areas mapped: http://www.parliament.uk/commons/lib/research/rp98/rp98–102.pdf).

The neighbourhood statistics website of the Office for National Statistics (http://www. neighbourhood.statistics.gov.uk/home.asp).

The General Register Office for Scotland website (http://www.gro-scotland.gov.uk/).

Results published from the Youth Cohort Study of England and Wales (http://www.statistics. gov.uk/STATBASE/Source.asp?vlnk=668&More=Y).

Publications of the Higher Education Council for England (in particular www.hefce.ac.uk/pubs/ hefce/2001/01 62.htm).

Seymour, J. (ed.), 2001, *Poverty in Plenty: A Human Development Report for the UK*, Earthscan Publications Limited, London, ISBN 1 85383 707 5 (http://www.earthscan.co.uk/asp/ bookdetails.asp?key=3142).

Work on worldwide trends published by The Townsend Centre for International Poverty Research at the University of Bristol (http://www.bris.ac.uk/poverty/) in association with UNICEF (http://www.unicef.org/sowc04/sowc04 special issues.html): Gordon, D., et al., 'The Distribution of Child Poverty in the Developing World: Report to UNICEF', Centre for International Poverty Research, University of Bristol, Bristol, 2003.

For further information type 'human geography of the UK' into the search box of google (http://www.google.com/). At the time of writing over one million web pages were linked to this phrase, ordered roughly by relevance to the subject. Don't be put off by how much is written about the UK – it remains a small island!

All web links were valid as of July 2004

Index

Indexed by Caroline Eley